Continuum Damage and Fracture Mechanics

Andreas Öchsner

Continuum Damage
and Fracture Mechanics

 Springer

Andreas Öchsner
Griffith School of Engineering
Griffith University (Gold Coast Campus)
Southport, QLD
Australia

ISBN 978-981-10-1323-2   ISBN 978-981-287-865-6   (eBook)
DOI 10.1007/978-981-287-865-6

Springer Singapore Heidelberg New York Dordrecht London
© Springer Science+Business Media Singapore 2016
Softcover re-print of the Hardcover 1st edition 2016

Printed on acid-free paper

Springer Science+Business Media Singapore Pte Ltd. is part of Springer Science+Business Media
(www.springer.com)

I am thankful to all the men and women who have dedicated their lives to keeping the rest of us safe and free enough to do things like read and write books.

Michael J. Mooney

Preface

This book results from the elective course 'Continuum Damage and Fracture Mechanics' (3510ENG) held at Griffith University, Australia, in the scope of the 'Bachelor of Engineering with Honours in Mechanical Engineering' degree program. This 13-week course comprises 2 hours of lectures and 2 hours of tutorials per week. Owing to the structure of the entire degree program, the course reviewed and extended in the first part, the classical theory of elastic and elasto-plastic material behavior. A thorough knowledge of these two topics is the essential prerequisite to cover the areas of damage and fracture mechanics. Thus, the second part of this course gives a first introduction to the treatment of damage and fracture in the scope of applied mechanics. Where possible, the one-dimensional case is first introduced and then in a following step generalized. This might be different to the more classical approach where first the most general case is derived and then simplified to special cases. In general, the requirements of mathematics are kept low and more challenging topics of damage and fracture mechanics are reserved for other courses, e.g., in the scope of a Master of Engineering program.

The supplementary problems within this book are taken from the tutorials offered to our students. A classical approach is here adopted (see Fig. 1) where a problem description is issued. The tutorial itself presents the major steps for the solution and gives tips and tricks. The final solution is as well available to the students. However, it is essential that students of engineering subjects try to find the way to the final solution on their own and not by simply looking at a full

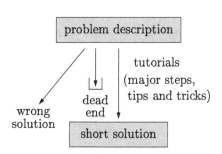

Fig. 1 Classical teaching and training approach in the scope of engineering education

step-by-step solution. Engineering skills and knowledge are acquired by actively getting trained to solve engineering problems. This is best achieved by taking an empty piece of paper to start the calculation and face the derivation difficulties from the very beginning.

While attempting to solve the problem, some students may reach a dead end, i.e., they do not have any further idea on how to proceed, or they may end up with a wrong solution. This is, from an educational point of view, not a major problem and should not be discouraging. Tutors and the lecturer will offer their help and provide guidance on how to overcome these when encountered.

This book should be understood as a first introduction to the topic and at least enable the reader to comprehend the basic ideas of the presented approaches of modeling in applied mechanics.

Gold Coast, Australia   Andreas Öchsner
June 2015

Acknowledgments

It is important to emphasize the contribution of many students which helped to finalize the content of this book. Their questions and comments during different lectures and tutorials helped to compile this book. The help and support of my tutors and Ph.D. students Leonhard Hitzler and Zia Javanbakht is gratefully acknowledged. Furthermore, I would like to express my sincere appreciation to the Springer-Verlag, especially to Dr. Christoph Baumann, for giving me the opportunity to realize this book. A professional publishing company with the right understanding was the prerequisite for completing this project. Finally, I would also like to thank my family for the understanding and patience during the preparation of this book.

Contents

Symbols and Abbreviations

Latin Symbols (Capital Letters)

A	Area, cross-sectional area, dimension
$\mathcal{A}$	Crack area
$\bar{A}$	Effective resisting area
C	Elasticity matrix
C^{elpl}	Elasto-plastic modulus matrix
D	Compliance, damage variable
$\boldsymbol{D}$	Compliance matrix
D^{pl}	Generalized plastic modulus
E	YOUNG's modulus
$\bar{E}$	Modulus of damaged material
F	Force, yield condition
F_{Q}	Conditional force
G	Shear modulus
$\mathcal{G}$	Energy release rate
$\mathcal{G}_{\text{c}}$	Critical energy release rate
H	Kinematic hardening modulus
I	Second moment of area
I_{p}	Polar second moment of area
I_i	Principal invariant of stress tensor $(i = 1, 2, 3)$
I_i^{o}	Principal invariant of spherical tensor $(i = 1, 2, 3)$
I_i'	Principal invariant of stress deviator $(i = 1, 2, 3)$
$\underline{I}_i$	Principal invariant of stress tensor $(i = 1, 2, 3)$ without symmetry of shear stress
J	J-integral
J_{c}	Critical value of the J-integral
J_i	Basic invariant of stress tensor $(i = 1, 2, 3)$
J_i^{o}	Basic invariant of spherical tensor $(i = 1, 2, 3)$
J_i'	Basic invariant of stress deviator $(i = 1, 2, 3)$

$\underline{J}_i$	Basic invariant of stress tensor ($i = 1, 2, 3$) without symmetry of shear stress
K	Bulk modulus
K_I	Stress intensity factor for mode I
K_{Ic}	Fracture toughness for mode I
K_{II}	Stress intensity factor for mode II
K_{IIc}	Fracture toughness for mode II
K_{III}	Stress intensity factor for mode III
K_{IIIc}	Fracture toughness for mode III
K_Q	Conditional fracture toughness
K_t	Stress concentration factor
$\boldsymbol{K}$	Stiffness matrix
L	Length
M	Moment
M_T	Torsional moment
$\boldsymbol{M}$	Mass matrix
Q	Shape factor for elliptical crack
R	Radius of curvature (necking), stress ratio (cyclic loading), electrical resistance
T	Temperature, torque
T_i	Traction vector
T_{kf}	Melting temperature
V	Crack mouth opening displacement, volume
$\bar{V}$	Effective resisting volume
W	Deformation energy
Y	Damage energy release rate
Y_c	Dimensionless parameter for stress intensity factor

Latin Symbols (Small Letters)

a	Crack size, dimension
Δa	Crack extension
b	Stress ratio, thickness
c	Dimension, speed of sound
d	Diameter
e	Geometrical dimension
$\boldsymbol{e}$	Unit vector
f	Void volume fraction
g	Geometrical dimension
h	Evolution function of hardening parameter, height
$\boldsymbol{h}$	Evolution function matrix of hardening parameters
k	Yield stress, spring constant
k_c	Compressive yield stress

k_s	Shear yield stress
k_t	Tensile yield stress
l_r	Remaining ligament
n_i	Components of unit vector normal to Γ
p	Pressure
q	Internal variable (hardening)
$\boldsymbol{q}$	Column matrix of hardening variables
r	Damage evolution parameter
$\boldsymbol{r}$	Plastic flow direction
s	Damage evolution parameter
s_{ij}	Deviatoric stress tensor
$\boldsymbol{s}$	Deviatoric stress vector
t	Wall thickness
u	Displacement
$\boldsymbol{u}$	Column matrix of displacements
w	Deformation energy per unit volume, geometrical dimension
w°	Volumetric deformation energy per unit volume
w^s	Distortional deformation energy per unit volume
x	Cartesian coordinate
y	Cartesian coordinate
z	Cartesian coordinate

Greek Symbols (Capital Letters)

Γ	Boundary
$\boldsymbol{\Phi}$	Column matrix of eigenmodes

Greek Symbols (Small Letters)

α	Back stress/kinematic hardening parameter
γ	Shear strain (engineering definition)
ε	Strain
$\bar{\varepsilon}$	Effective strain
$\varepsilon_{\text{eff}}^{\text{pl}}$	Effective plastic strain (equivalent plastic strain)
$\varepsilon_t^{\text{init}}$	Initial tensile yield strain
ε_t^{f}	Tensile strain at fracture
ε_v	Volumetric strain
$\varepsilon_v^{\text{pl}}$	Volumetric plastic strain
$\boldsymbol{\varepsilon}$	Column matrix of strain components
η	Factor
ϑ_L	Total twist angle
θ	Lode angle
κ	Isotropic hardening parameter, Factor

λ	LAMÉ's constant, consistency parameter (cf. Plasticity)
μ	LAMÉ's constant
ν	POISSON's ratio
ξ	HAIGH–WESTERGAARD coordinate
ρ	HAIGH–WESTERGAARD coordinate
σ	Stress
$\bar{\sigma}$	Effective stress
σ_c	Critical stress
σ_{eff}	Effective stress (equivalent stress)
σ_m	Mean normal stress
σ_i	Principal stresses ($i = 1, 2, 3$)
σ_{ij}	Stress tensor
σ_{ij}^o	Spherical (hydrostatic) stress tensor
$\sigma_t^{\max}$	Ultimate tensile strength
σ_t^f	Tensile fracture stress
σ	Column matrix of stress components
τ	Shear stress
φ	Angle
ω	Eigenfrequency

Mathematical Symbols

$(\ldots)^{\mathrm{T}}$	Transpose
$\times$	Multiplication sign (used where essential)
$\det(\ldots)$	Determinant
$\dim(\ldots)$	Dimension
$\text{sgn}(\ldots)$	Signum (sign) function
IR	Set of real numbers

Special Matrices and Vectors

$\mathbf{I}$	Identity matrix (diagonal matrix)

Indices, Superscripted

$\ldots^{\text{el}}$	Elastic
$\ldots^{\text{init}}$	Initial
$\ldots^f$	Failure
$\ldots^{\max}$	Maximum
$\ldots^{\text{pl}}$	Plastic
$\ldots^r$	Rupture

Indices, Subscripted

$\cdots_0$	Initial
$\cdots_c$	Compression, critical, cylindrical
$\cdots_D$	Damaged
$\cdots_{ext}$	External
$\cdots_H$	Hoop
$\cdots_i$	Initial
$\cdots_{int}$	Internal
$\cdots_L$	Longitudinal
$\cdots_{max}$	Maximum
$\cdots_{nom}$	Nominal
$\cdots_s$	Hemispherical
$\cdots_t$	Tensile
$\cdots_T$	Torsional
$\cdots_{tr}$	True

Abbreviations

1D	One-dimensional
ASTM	American Society for Testing and Materials
BEM	Boundary element method
CMOD	Crack mouth opening displacement
CT	Compact tension
DCB	Double cantilever beam
EDX	Energy dispersive X-ray spectroscopy
FDM	Finite difference method
FEM	Finite element method
FM	Fracture mechanics
FVM	Finite volume method
PAT	Principal axis transformation
RVE	Representative volume element
SEM	Scanning electron microscope

Some Standard Abbreviations

ca.	About, approximately (from Latin 'circa')
cf.	Compare (from Latin 'confer')
ead.	The same (female) (from Latin 'eadem')
e.g.	For example (from Latin 'exempli gratia')
et al.	And others (from Latin 'et alii')
et seq.	And what follows (from Latin 'et sequens')
etc.	And others (from Latin 'et cetera')
i.a.	Among other things (from Latin 'inter alia')

ibid.	In the same place (the same), used in citations (from Latin 'ibidem')
id.	The same (male) (from Latin 'idem')
i.e.	That is (from Latin 'id est')
loc. cit.	In the place cited (from Latin 'loco citato')
N.N.	Unknown name, used as a placeholder for unknown names (from Latin 'nomen nescio')
op. cit.	In the work cited (from Latin 'opere citato')
pp.	Pages
q.e.d.	Which had to be demonstrated (from Latin 'quod erat demonstrandum')
viz.	Namely, precisely (from Latin 'videlicet')
vs.	Against (from Latin 'versus')

Chapter 1
Introduction

Abstract The first chapter classifies the content as well as the focus of this textbook. Based on the classical tensile test in structural mechanics, the engineering stress-strain diagram is introduced and the various theories covered in this textbook are described based on the material behavior on the microscale.

Let us consider in the following a uniaxial tensile test whose idealized specimens are schematically shown in Figs. 1.1 and 1.2. The original dimensions of the specimens are characterized by the initial cross-sectional area A_0 and length L. This specimen is now elongated in a universal testing machine and its length increases to $L + \Delta L$. In the case of a specimen made of a classical engineering material, the cross-sectional area reduces to $A_0 - \Delta A$. This phenomenon can be described based on POISSON's ratio.[1] Thus, we have a uniaxial stress state, i.e. just the stress component in the loading direction, and a multiaxial strain state due to the specimen's contraction.

The most common specimen shapes that have been used for tensile testing are cylindrical (Fig. 1.1) or rectangular (Fig. 1.2). For some materials, the raw material might be only available in the forms of rods or plates, which results obviously in one of the mentioned shapes. Round tensile specimens allow a simpler observation and evaluation of the specimen's geometry at larger strains while a rectangular specimens allows a simple application of strain gauges, e.g. a 0/90° biaxial configuration to measure the longitudinal and transversal strains.

During the tensile test, the force F is normally recorded by a load cell attached to the movable or fixed cross-head of the machine. If this force is divided by the *initial* cross-sectional area A_0, the engineering stress is obtained as:

$$\sigma = \frac{F}{A_0}. \tag{1.1}$$

The deformation or elongation of the specimen can be measured, for example, by an external extensometer which should be directly attached to the specimen. These devices are either realized as strain gauge or inductive extensometers.[2] Any

---

[1] Siméon Denis POISSON (1781–1840), French mathematician, geometer, and physicist.

[2] More modern devises are contactless laser or video based extensometers.

© Springer Science+Business Media Singapore 2016
A. Öchsner, *Continuum Damage and Fracture Mechanics*,
DOI 10.1007/978-981-287-865-6_1

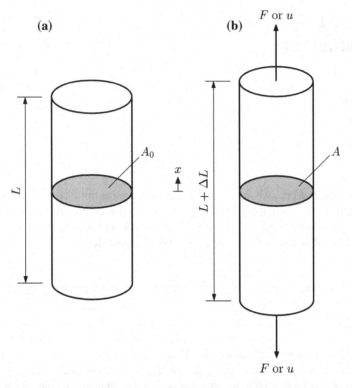

Fig. 1.1 Schematic representation of **a** an unloaded and **b** a loaded round tensile specimen. The load is applied either by a force F or a displacement u (the grip ends are not shown for simplicity)

measurements based on the movement of the machine's cross-head must be avoided since they do not guarantee an accurate determination of the specimen's behavior. The definition of strain is given in its simplest form as elongation over initial length and the engineering strain can be calculated as:

$$\varepsilon = \frac{\Delta L}{L}.$$
(1.2)

In the case of larger deformations and strains, it is also common to calculate the true stress, i.e.

$$\sigma_{\text{tr}} = \frac{F}{A},$$
(1.3)

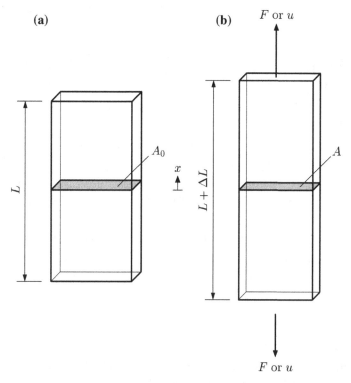

(a)

(b) F or u

A_0

A

L

x

$L + \Delta L$

F or u

Fig. 1.2 Schematic representation of **a** an unloaded and **b** a loaded flat tensile specimen. The load is applied either by a force F or a displacement u (the grip ends are not shown for simplicity)

where the force F is divided by the actual cross-sectional area A. The true strain is the natural logarithm of the actual length $L + \Delta L$ over the initial length L:

$$\varepsilon_{\mathrm{tr}} = \ln\left(\frac{L + \Delta L}{L}\right) = \ln(1 + \varepsilon) . \tag{1.4}$$

The true plastic strain is obtained by subtracting the elastic strain part from the total value as follows:

$$\varepsilon_{\mathrm{tr}}^{\mathrm{pl}} = \ln(1 + \varepsilon) - \varepsilon_{\mathrm{tr}}^{\mathrm{el}} = \ln(1 + \varepsilon) - \frac{\sigma_{\mathrm{tr}}}{E} . \tag{1.5}$$

A typical engineering stress-strain diagram of a ductile aluminium alloy is shown in Fig. 1.3 where the distinct plastic region can be observed.

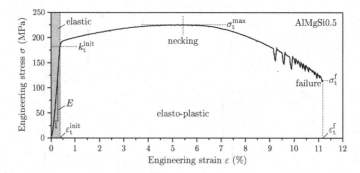

Fig. 1.3 Typical engineering stress-strain diagram of a ductile aluminium alloy (AlMgSi0.5). Several characteristic material parameters and regions are indicated. Adapted from [42]

This diagram allows to extract several characteristic material parameters such as:

- YOUNG's[3] modulus E,
- initial tensile yield stress k_t^{init},
- ultimate tensile strength σ_t^{max},
- tensile fracture stress σ_t^f,
- initial tensile yield strain ε_t^{init},
- tensile strain at fracture ε_t^f.

The difference between the engineering and true stress is schematically shown in Fig. 1.4. It can be seen that a significant difference occurs after the necking point, i.e. the true stress is monotonically increasing whereas the engineering stress is decreasing.

Looking at the microstructure of a material and its evolution during the tensile test (see Fig. 1.5), different stages can be distinguished. The evaluation of micrographs of a ductile aluminium alloy (AlMgSi0.5) obtained from a scanning electron microscope (SEM) shows that the ductile damage occurs simultaneously with plastic deformation larger than a certain threshold. In the initial state or at deformations before necking, a

Fig. 1.4 Schematic difference between engineering σ and true stress σ_{tr} calculation

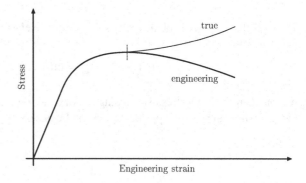

---

[3]Thomas YOUNG (1773–1829), English polymath.

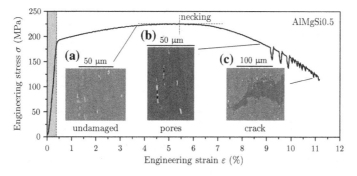

Fig. 1.5 Typical engineering stress-strain diagram of a ductile aluminium alloy and corresponding SEM micrographs: **a** region of uniform deformation showing precipitates, **b** after necking with formation of pores near the precipitates, **c** close to fracture with internal crack. Adapted from [42]

Fig. 1.6 SEM micrograph (longitudinal section) of the necked region of a round tensile sample. Unloaded shortly before fracture, showing the formed macro-crack. Adapted from [21]

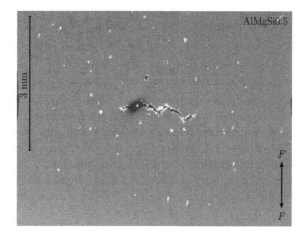

classical micrograph is obtained which shows the matrix and precipitates as smaller light regions, see Fig. 1.5a. At larger strains after necking, it can be observed that the usually brittle precipitates can break or separate from the matrix, both resulting in the formation of pores. These voids join together under tensile load and grow, see Fig. 1.5b. This mechanism leads to microcracking and subsequently to the formation of a macro-crack which ultimately results in the failure of the material, see Fig. 1.5c.

A SEM micrograph (longitudinal section) of the necked region of a round tensile sample is shown in Fig. 1.6 where the minimum cross section can be seen at the very left and right-hand side of the picture. The internal crack is located in the center of the specimen which corresponds to the location of the highest stress triaxiality[4] and highest plastic strain.

---

[4]The stress triaxiality ratio is the fraction between the mean and the equivalent VON MISES stress and describes how multiaxial a stress state is. In the case of a pure uniaxial stress state, the stress triaxiality ratio takes a value of one-third [67].

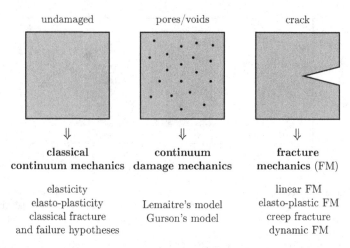

Fig. 1.7 Differentiation of the various theories

At the end of this chapter, let us have a look on the different theories covered in the book (see Fig. 1.7). We will start with classical continuum mechanics which treats a material continuously distributed and filling the entire space it occupies. The material is considered homogeneous without—at the considered length scale—any microstructure, defects, pores, etc. Continuum damage mechanics considers the effect of pores, voids and microcracks on the mechanical properties but still considers the material as homogeneous. Fracture mechanics considers discontinuities in the material in the form of cracks and their influence on the mechanical properties and final failure.

1.1 Supplementary Problems

1.1 Knowledge questions

- Name one advantage of a flat tensile specimen when compared to a round configuration.
- Explain the difference between the engineering and true stress.
- Name the material parameters which can be extracted from an engineering stress-strain diagram of a ductile metal.
- Sketch the engineering stress-strain diagram of a ductile and a brittle metal.
- Explain the major ideas of (a) classical continuum mechanics, (b) damage mechanics and (c) fracture mechanics.

1.2 Approximation of the true stress
Derive under the assumption that the volume remains constant in the plastic range a simplified expression for the true stress σ_{tr} which depends only on the applied force, the initial cross-sectional area and the engineering strain.

1.3 Approximation of the true strain
Show that the expression for the true strain given in Eq. (1.4) converges towards the engineering strain expression for small strain values.

1.4 Engineering versus true stress and strain representation
The engineering stress-strain diagram of a ductile metal can be approximated by the following two equations which describe the pure elastic and the elasto-plastic range:

$$
\begin{aligned}
\sigma(\varepsilon) &= E \times \varepsilon && \text{for} && 0 \le \varepsilon \le 0.0025\,, \\
\sigma(\varepsilon) &= a\varepsilon^2 + b\varepsilon + c && \text{for} && 0.0025 \le \varepsilon \le 0.1\,,
\end{aligned}
\tag{1.6}
$$

where $E = 72000$ MPa, $a = -17728.532$ MPa, $b = 1772.853$ MPa and $c = 175.679$ MPa. Draw the engineering/true stress—engineering strain diagram and the true stress-engineering/true strain diagram. Illustrate the difference between the engineering and true strain.

Chapter 2
Elastic Material Behavior

Abstract This chapter reviews the basics of elastic material behavior. Starting from simple load cases, i.e. uniaxial tension, pure shear and hydrostatic compression, basic material parameters are derived from experimental results. The next part presents the three-dimensional constitutive law for isotropic and linear-elastic material behavior. The chapter closes with two important cases of two-dimensional formulations, i.e. the plane stress and the plane strain case.

2.1 Simple Load Cases

2.1.1 Uniaxial Tensile Test

Let us consider in the following a tensile sample of length L and diameter d which is loaded by a linearly increasing external force F, see Fig. 2.1. In addition to the applied force F, the elongation ΔL and change in diameter Δd is recorded during the test.

Using the definitions of engineering stress and strain as given in Eqs. (1.1) and (1.2), i.e.

$$\sigma = \frac{F}{A_0},\tag{2.1}$$

and the longitudinal strain

$$\varepsilon_x = \frac{\Delta L}{L},\tag{2.2}$$

the engineering stress-strain diagram in the linear-elastic region results as shown in Fig. 2.2a. For most of the engineering materials, a straight line is observed and its slope is equal to YOUNG's modulus:

$$E = \frac{\Delta \sigma}{\Delta \varepsilon_x}.\tag{2.3}$$

© Springer Science+Business Media Singapore 2016
A. Öchsner, *Continuum Damage and Fracture Mechanics*,
DOI 10.1007/978-981-287-865-6_2

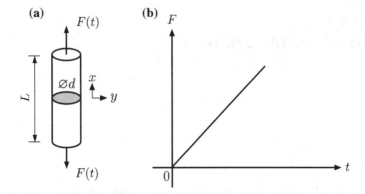

Fig. 2.1 Uniaxial tensile test: **a** idealized specimen and **b** force-time distribution

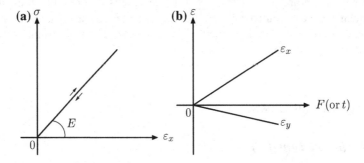

Fig. 2.2 Linear-elastic material behaviour: **a** engineering stress-strain diagram; **b** course of longitudinal and transversal strain

If the specimen is elongated, the cross-sectional area of all classical engineering materials is reduced and the transversal strain is defined as:

$$\varepsilon_y = \frac{\Delta d}{d}. \tag{2.4}$$

The course of the longitudinal and transversal strain is shown in Fig. 2.2b.

Relating the transversal to the longitudinal strain defines POISSON's ratio:

$$\nu = -\frac{\varepsilon_y}{\varepsilon_x}. \tag{2.5}$$

Mechanical and physical reference values[1] of some typical engineering materials are summarised in Table 2.1.

---

[1] Conversion from GPa to MPa: value times 1000; conversion from g × cm^{-3} to kg × m^{-3}: value times 1000; 1 Pa = 1 N/m^2.

Table 2.1 Reference values of typical engineering materials (all values are given near room temperature)

Material	E in GPa	ν in –	Density in $g \times cm^{-3}$
Aluminium	70	0.33	2.700
Steel	210	0.30	7.874
Magnesium	45	0.35	1.738
Titanium	110	0.34	4.506

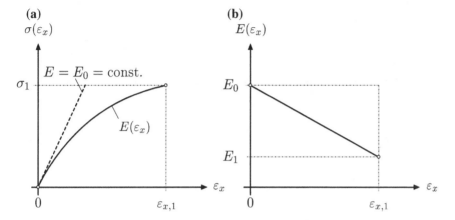

Fig. 2.3 a Nonlinear-elastic stress-strain diagram; **b** strain-dependent modulus of elasticity

Some materials show an elastic material behavior, but the course of the stress-strain diagram is nonlinear, see Fig. 2.3a.

In such a case, the YOUNG's modulus is not constant, i.e. a function of the acting strain. Thus, Eq. (2.3) must be modified to a differential definition as:

$$E(\varepsilon_x) = \frac{d\sigma}{d\varepsilon_x}. \tag{2.6}$$

2.1.2 Pure Shear Test

Let us consider in the following a cylindrical specimen of length L and diameter d which is loaded by a linearly increasing external torsional moment M_T, see Fig. 2.4. In addition to the applied moment M_T, the total twist angle ϑ_L at the end of the specimen ($x = L$) is recorded during the test. This test can be realized in a specific torsion testing machine or in a classical universal testing machine under usage of a special jig which creates the load by a lever arm.

Plotting the torsional moment over the twist angle results in a straight line as shown in Fig. 2.5a. However, these quantities are still dependent on the size of the specimen, similar to the force and elongation in the case of the tensile test.

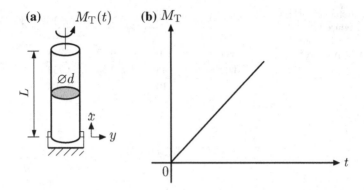

Fig. 2.4 Pure shear test: **a** idealized specimen and **b** moment-time distribution

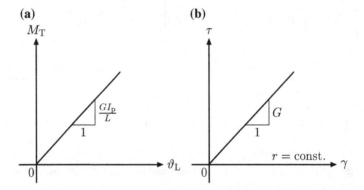

Fig. 2.5 Pure shear test: **a** moment—twisting angle distribution **b** shear stress—shear strain diagram

Normalizing these quantities results in a shear stress versus shear strain diagram (see Fig. 2.5b) and the so-called shear modulus can be identified as the slope:

$$G = \left. \frac{\tau}{\gamma} \right|_{r=\text{const.}} \tag{2.7}$$

In the case of a cylindrical specimen which is clamped at one end and the other end is loaded by a torsional moment, the following relation can be derived [24, 64][2]:

$$G = \frac{M_T L}{I_p \vartheta_L}, \tag{2.8}$$

---

[2]See supplementary Problem 2.4.

where I_p is the polar second moment of area (in the case of a cylinder: $I_p = \frac{\pi d^4}{32}$) and ϑ_L is the total twist angle at the end of the cylindrical specimen. Equation (2.8) can be written as

$$G = \frac{\frac{M_T}{I_p}}{\frac{\vartheta_L}{L}},$$ (2.9)

which corresponds to the formulation of the YOUNG's modulus: $E = \frac{F/A}{\Delta L/L}$.

2.1.3 Hydrostatic Compression Test

Let us consider in the following a cubic sample which is loaded from all three sides by linear increasing external compressive forces $F (\sim \sigma)$, see Fig. 2.6. In addition to the applied forces F, the shortening is recorded during the test. These quantities can be easily converted into a compressive stress and a compressive (axial) strain which is, due to the symmetry of the problem in all three directions (x, y, z), the same. Such a test can be realized in a triaxial testing machine, i.e. a machine which allows simultaneous loading of specimens along three axes perpendicular to each other, or by enclosing specimens in a hydraulic cylinder where a fluid pressure is applied by a piston.

Plotting the mean stress over the volumetric strain,[3] a linear dependency is obtained which allows the definition of the bulk modulus.

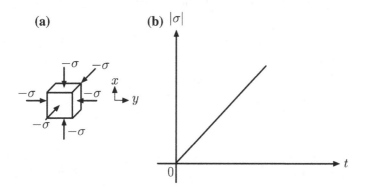

Fig. 2.6 Hydrostatic compression test: **a** idealized specimen and **b** stress-time distribution

---

[3] See supplementary Problems 2.2 and 2.6.

Fig. 2.7 Mean normal stress
versus volumetric strain

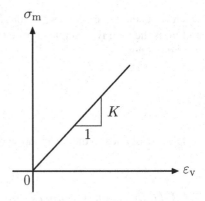

The bulk modulus K is then defined as the ratio between the mean normal stress (hydrostatic stress) and the corresponding volume change [15] (Fig. 2.7):

$$K = \frac{\sigma_{\mathrm{m}}}{\varepsilon_{\mathrm{v}}} = \frac{\frac{1}{3}\sigma_{kk}}{\varepsilon_{kk}}.$$

(2.10)

2.2 Three-Dimensional Hooke's Law

Let us first look on the general concept of continuum mechanical modelling of material behavior, see Fig. 2.8. The equilibrium equation relates the external forces to the internal reactions and is a measure for the loading of the material. The kinematics

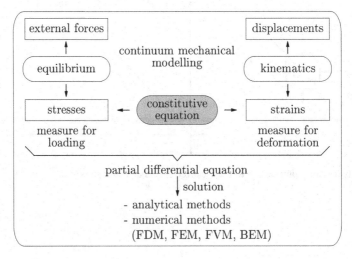

Fig. 2.8 Continuum mechanical modelling

equation relates the displacements to the strains and is a measure for the deformation of the body. The constitutive equation relates stress and strain. The following section treats the simplest three-dimensional case, i.e. HOOKE's law for isotropic linear-elastic material behavior.

Combining these three equations, i.e. equilibrium, kinematics and constitution, results in a partial differential equation which describes the entire problem at any location of the material domain. For simple geometries (e.g. beams or rods), analytical solutions can be derived which offer an exact description of the problem under the given assumptions and simplifications. For more complicated cases, approximate solutions can be obtained based on numerical techniques such as the finite element method (FEM), see [43, 44].

Let us look in the following at a three-dimensional body which is sufficiently supported and loaded as shown in Fig. 2.9a. Isolating a differential volume element, the internal stresses—which are in equilibrium with the external loads—occur, see Fig. 2.9b.

The nine stresses[4] σ_{ij} are the components of the second-order stress tensor and are arranged in the following way:

$$\sigma_{ij} = \begin{bmatrix} \sigma_{xx} & \sigma_{xy} & \sigma_{xz} \\ \sigma_{yx} & \sigma_{yy} & \sigma_{yz} \\ \sigma_{zx} & \sigma_{zy} & \sigma_{zz} \end{bmatrix} \rightarrow \sigma = \{\sigma_x \ \sigma_y \ \sigma_z \ \sigma_{xy} \ \sigma_{yz} \ \sigma_{xz}\}^T, \qquad (2.11)$$

where σ is the column matrix of the stress components which contains only the independent variables (engineering notation). This reduces the required disk space to store the stress values. In a similar way, the corresponding strains are arranged:

$$\varepsilon_{ij} = \begin{bmatrix} \varepsilon_{xx} & \varepsilon_{xy} & \varepsilon_{xz} \\ \varepsilon_{yx} & \varepsilon_{yy} & \varepsilon_{yz} \\ \varepsilon_{zx} & \varepsilon_{zy} & \varepsilon_{zz} \end{bmatrix} \rightarrow \varepsilon = \{\varepsilon_x \ \varepsilon_y \ \varepsilon_z \ 2\varepsilon_{xy} \ 2\varepsilon_{yz} \ 2\varepsilon_{xz}\}^T. \qquad (2.12)$$

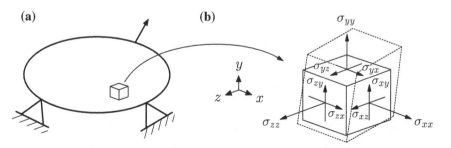

(a) **(b)**

Fig. 2.9 **a** Three-dimensional body under arbitrary load; **b** differential volume element with acting stresses

[4]The first index i indicates that the stress acts on a plane normal to the i-axis and the second index j denotes the direction in which the stress acts.

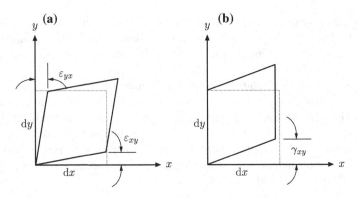

Fig. 2.10 Definition of shear strain: **a** tensor definition as $\varepsilon_{xy} \approx \partial u_y / \partial x$ and $\varepsilon_{yx} \approx \partial u_x / \partial y$; **b** engineering definition as the total $\gamma_{xy} \approx \partial u_y / \partial x + \partial u_x / \partial y$

It should be noted here that there are two definitions for the shear strain in use: The tensor definition ε_{ij} (for $i \neq j$) and the engineering shear strain $\gamma_{ij} = 2\varepsilon_{ij}$ (for $i \neq j$), see Fig. 2.10.

In the following description, the tensor notation is abandoned and the engineering notation (VOIGT notation) is now consistently introduced and observed, i.e. the components of the second-order stress tensor σ_{ij} and the strain tensor ε_{ij} are arranged into column matrices σ and ε. Since the stress and strain tensor are symmetric (e.g. $\sigma_{ij} = \sigma_{ji}$), it is more convenient and economic to store only the six independent components in a single column matrix. The constitutive equation which relates the stresses and strains (see Fig. 2.8) can be generally expressed as

$$\sigma = C\varepsilon \,, \tag{2.13}$$

or

$$\varepsilon = D\sigma \,, \tag{2.14}$$

where C is the so-called elasticity matrix and D the so-called elastic compliance matrix. Let us assume the following simplifications for the material under consideration:

- linear elastic,
- isotropic,
- homogeneous,
- isothermal conditions.

In extension to HOOKE's law ($\sigma = E\varepsilon$) of the year 1678, the following generalized formulation based on the engineering constants YOUNG's modulus E and POISSON's ratio ν can be given:

$$
\begin{bmatrix}
\sigma_x \\
\sigma_y \\
\sigma_z \\
\sigma_{xy} \\
\sigma_{yz} \\
\sigma_{xz}
\end{bmatrix}
=
\frac{E}{(1+\nu)(1-2\nu)}
\begin{bmatrix}
1-\nu & \nu & \nu & 0 & 0 & 0 \\
\nu & 1-\nu & \nu & 0 & 0 & 0 \\
\nu & \nu & 1-\nu & 0 & 0 & 0 \\
0 & 0 & 0 & \frac{1-2\nu}{2} & 0 & 0 \\
0 & 0 & 0 & 0 & \frac{1-2\nu}{2} & 0 \\
0 & 0 & 0 & 0 & 0 & \frac{1-2\nu}{2}
\end{bmatrix}
\begin{bmatrix}
\varepsilon_x \\
\varepsilon_y \\
\varepsilon_z \\
2\varepsilon_{xy} \\
2\varepsilon_{yz} \\
2\varepsilon_{xz}
\end{bmatrix}.
\tag{2.15}
$$

Rearranging the elastic stiffness form given in Eq. (2.15) for the strains gives the elastic compliance form:

$$
\begin{bmatrix}
\varepsilon_x \\
\varepsilon_y \\
\varepsilon_z \\
2\varepsilon_{xy} \\
2\varepsilon_{yz} \\
2\varepsilon_{xz}
\end{bmatrix}
=
\frac{1}{E}
\begin{bmatrix}
1 & -\nu & -\nu & 0 & 0 & 0 \\
-\nu & 1 & -\nu & 0 & 0 & 0 \\
-\nu & -\nu & 1 & 0 & 0 & 0 \\
0 & 0 & 0 & 2(1+\nu) & 0 & 0 \\
0 & 0 & 0 & 0 & 2(1+\nu) & 0 \\
0 & 0 & 0 & 0 & 0 & 2(1+\nu)
\end{bmatrix}
\begin{bmatrix}
\sigma_x \\
\sigma_y \\
\sigma_z \\
\sigma_{xy} \\
\sigma_{yz} \\
\sigma_{xz}
\end{bmatrix}.
\tag{2.16}
$$

In a simple tension test, the only nonzero stress component σ_x causes axial strain ε_x and transverse strains $\varepsilon_y = \varepsilon_z$. Thus, Eqs. (2.15) and (2.16) yield

$$
\varepsilon_x = \frac{\sigma_x}{E} \qquad \text{and} \qquad \varepsilon_y = -\nu\varepsilon_x = -\frac{\nu\sigma_x}{E}.
\tag{2.17}
$$

By using Eq. (2.17), one can experimentally determine the elastic constants, i.e. YOUNG's modulus E and POISSON's ratio ν, from a uniaxial tension or compression test, see Chap. 1 and Sect. 2.1.1.

Replacing the YOUNG's modulus E and POISSON's ν ratio by the shear modulus G and bulk modulus K gives:

$$
\begin{bmatrix}
\sigma_x \\
\sigma_y \\
\sigma_z \\
\sigma_{xy} \\
\sigma_{yz} \\
\sigma_{xz}
\end{bmatrix}
=
\begin{bmatrix}
K+\frac{4}{3}G & K-\frac{2}{3}G & K-\frac{2}{3}G & 0 & 0 & 0 \\
K-\frac{2}{3}G & K+\frac{4}{3}G & K-\frac{2}{3}G & 0 & 0 & 0 \\
K-\frac{2}{3}G & K-\frac{2}{3}G & K+\frac{4}{3}G & 0 & 0 & 0 \\
0 & 0 & 0 & G & 0 & 0 \\
0 & 0 & 0 & 0 & G & 0 \\
0 & 0 & 0 & 0 & 0 & G
\end{bmatrix}
\begin{bmatrix}
\varepsilon_x \\
\varepsilon_y \\
\varepsilon_z \\
2\varepsilon_{xy} \\
2\varepsilon_{yz} \\
2\varepsilon_{xz}
\end{bmatrix},
\tag{2.18}
$$

or as the elastic compliance form:

$$
\begin{bmatrix}
\varepsilon_x \\
\varepsilon_y \\
\varepsilon_z \\
2\varepsilon_{xy} \\
2\varepsilon_{yz} \\
2\varepsilon_{xz}
\end{bmatrix}
=
\frac{1}{18KG}
\begin{bmatrix}
6K+2G & -3K+2G & -3K+2G & 0 & 0 & 0 \\
-3K+2G & 6K+2G & -3K+2G & 0 & 0 & 0 \\
-3K+2G & -3K+2G & 6K+2G & 0 & 0 & 0 \\
0 & 0 & 0 & 18K & 0 & 0 \\
0 & 0 & 0 & 0 & 18K & 0 \\
0 & 0 & 0 & 0 & 0 & 18K
\end{bmatrix}
\begin{bmatrix}
\sigma_x \\
\sigma_y \\
\sigma_z \\
\sigma_{xy} \\
\sigma_{yz} \\
\sigma_{xz}
\end{bmatrix}.
\tag{2.19}
$$

Table 2.2 Conversion of elastic constants λ, μ LAMÉ's constants; K bulk modulus; G shear modulus; E YOUNG's modulus; ν POISSON's ratio, [14]

	λ, μ	E, ν	μ, ν	E, μ	K, ν	G, ν	K, G
λ	λ	$\dfrac{\nu E}{(1+\nu)(1-2\nu)}$	$\dfrac{2\mu\nu}{1-2\nu}$	$\dfrac{\mu(E-2\mu)}{3\mu-E}$	$\dfrac{3K\nu}{1+\nu}$	$\dfrac{2G\nu}{1-2\nu}$	$K-\dfrac{2G}{3}$
μ	μ	$\dfrac{E}{2(1+\nu)}$	μ	μ	$\dfrac{3K(1-2\nu)}{2(1+\nu)}$	G	G
K	$\lambda+\tfrac{2}{3}\mu$	$\dfrac{E}{3(1-2\nu)}$	$\dfrac{2\mu(1+\nu)}{3(1-2\nu)}$	$\dfrac{\mu E}{3(3\mu-E)}$	K	$\dfrac{2G(1+\nu)}{3(1-2\nu)}$	K
E	$\dfrac{\mu(3\lambda+2\mu)}{\lambda+\mu}$	E	$2\mu(1+\nu)$	E	$3K(1-2\nu)$	$2G(1+\nu)$	$\dfrac{9KG}{3K+G}$
ν	$\dfrac{\lambda}{2(\lambda+\mu)}$	ν	ν	$\dfrac{E}{2\mu}-1$	ν	ν	$\dfrac{3K-2G}{2(3K+G)}$
G	μ	$\dfrac{E}{2(1+\nu)}$	μ	μ	$\dfrac{3K(1-2\nu)}{2(1+\nu)}$	G	G

Let us note at the end of this section that the general characteristic of HOOKE's law in the form of, for example, Eqs. (2.15) and (2.16) is that two independent material parameters are used. In addition to the YOUNG's modulus E and POISSON's ratio ν, other elastic parameters can be used to form the set of two independent material parameters, and the following Table 2.2 summarizes the conversion between the common material parameters.

2.3 Plane Stress and Plane Strain Case

2.3.1 Plane Stress Case

The two-dimensional plane stress case ($\sigma_z = \sigma_{yz} = \sigma_{xz} = 0, \varepsilon_z \neq 0$) shown in Fig. 2.11 is commonly used for the analysis of thin, flat plates loaded in the plane of the plate (x-y plane).

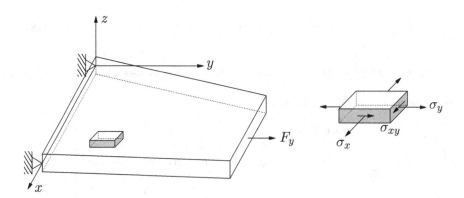

Fig. 2.11 Two-dimensional problem: plane stress

By imposing the condition $\sigma_z = 0$, we get from Eq. (2.15)

$$\sigma_z = \frac{E}{(1+\nu)(1-2\nu)} \cdot \left(\nu(\varepsilon_x + \varepsilon_y) + (1-\nu)\varepsilon_z\right) = 0 \qquad (2.20)$$

and

$$\varepsilon_z = -\frac{\nu}{1-\nu} \cdot \left(\varepsilon_x + \varepsilon_y\right). \qquad (2.21)$$

Substituting now ε_z into Eq. (2.15) and respecting $\sigma_z = \sigma_{yz} = \sigma_{xz} = 0$, we obtain the following form of HOOKE's law:

$$\begin{bmatrix} \sigma_x \\ \sigma_y \\ \sigma_{xy} \end{bmatrix} = \frac{E}{1-\nu^2} \begin{bmatrix} 1 & \nu & 0 \\ \nu & 1 & 0 \\ 0 & 0 & \frac{1-\nu}{2} \end{bmatrix} \begin{bmatrix} \varepsilon_x \\ \varepsilon_y \\ 2\varepsilon_{xy} \end{bmatrix}. \qquad (2.22)$$

Rearranging the elastic stiffness form given in Eq. (2.22) for the strains gives the elastic compliance form

$$\begin{bmatrix} \varepsilon_x \\ \varepsilon_y \\ 2\varepsilon_{xy} \end{bmatrix} = \frac{1}{E} \begin{bmatrix} 1 & -\nu & 0 \\ -\nu & 1 & 0 \\ 0 & 0 & 2(\nu+1) \end{bmatrix} \begin{bmatrix} \sigma_x \\ \sigma_y \\ \sigma_{xy} \end{bmatrix}. \qquad (2.23)$$

The general characteristic of plane HOOKE's law in the form of Eqs. (2.22) and (2.23) is that two independent material parameters are used.

It should be finally noted that the thickness strain ε_z can be obtained based on the two in-plane normal strains ε_x and ε_y as:

$$\varepsilon_z = -\frac{\nu}{1-\nu} \cdot \left(\varepsilon_x + \varepsilon_y\right). \qquad (2.24)$$

The last equation can be derived from the tree-dimensional formulation, see Sect. 2.2.

2.3.2 Plane Strain Case

The two-dimensional plane strain case ($\varepsilon_z = \varepsilon_{yz} = \varepsilon_{xz} = 0$) shown in Fig. 2.12 is commonly used for the analysis of elongated prismatic bodies of uniform cross section subjected to uniform loading along their radial axis but without any component in direction of the z-axis (e. g. pressure p_1 and p_2), such as in the case of tunnels, soil slopes, and retaining walls.

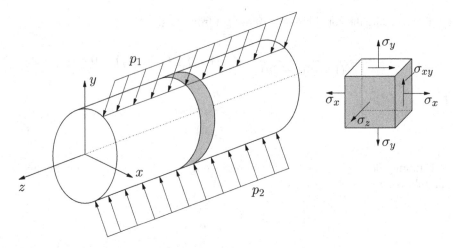

Fig. 2.12 Two-dimensional problem: plane strain

Considering that the stress components σ_{yz} and σ_{xz} are zero, Eq. (2.15) can be directly reduced to the plane strain form:

$$\begin{bmatrix} \sigma_x \\ \sigma_y \\ \sigma_{xy} \end{bmatrix} = \frac{E}{(1+\nu)(1-2\nu)} \begin{bmatrix} 1-\nu & \nu & 0 \\ \nu & 1-\nu & 0 \\ 0 & 0 & \frac{1-2\nu}{2} \end{bmatrix} \cdot \begin{bmatrix} \varepsilon_x \\ \varepsilon_y \\ 2\,\varepsilon_{xy} \end{bmatrix}. \tag{2.25}$$

Rearranging the elastic stiffness form given in Eq. (2.25) for the strains gives the elastic compliance form

$$\begin{bmatrix} \varepsilon_x \\ \varepsilon_y \\ 2\,\varepsilon_{xy} \end{bmatrix} = \frac{1-\nu^2}{E} \begin{bmatrix} 1 & -\frac{\nu}{1-\nu} & 0 \\ -\frac{\nu}{1-\nu} & 1 & 0 \\ 0 & 0 & \frac{2}{1-\nu} \end{bmatrix} \begin{bmatrix} \sigma_x \\ \sigma_y \\ \sigma_{xy} \end{bmatrix}. \tag{2.26}$$

By imposing the condition $\varepsilon_z = 0$, we get from Eq. (2.16)

$$\varepsilon_z = \frac{1}{E} \cdot \left(-\nu(\sigma_x + \sigma_y) + \sigma_z \right) = 0, \tag{2.27}$$

and the stress component σ_z can be obtained based on the two in-plane normal stresses σ_x and σ_y as:

$$\sigma_z = \nu(\sigma_x + \sigma_y). \tag{2.28}$$

2.4 Supplementary Problems

2.1 Knowledge questions

- How many material parameters are required for one-dimensional HOOKE's law?
- State the one-dimensional HOOKE's law for a pure *normal* stress and *normal* strain state.
- State the one-dimensional HOOKE's law for a pure *shear* stress and shear strain state.
- How many material parameters are required for the three-dimensional HOOKE's law?
- Explain the assumptions for an (a) 'isotropic' and (b) 'homogeneous' material.
- State the major characteristic of an *elastic* material.
- State approximate values for the YOUNG's modulus of (a) steel and (b) aluminium alloys.
- Which quantities does the constitutive law relate to?
- How many independent stress components occur in a general three-dimensional stress state?
- How many independent stress and strain components are acting in a plane stress state?
- How many independent stress and strain components are acting in a plane strain state?

2.2 Approximation of the volume change in a hydrostatic compression test
Consider a cuboid with dimensions $a \times b \times c$ which is compressed in a hydrostatic compression test. Derive an approximate equation for the volume change $\frac{\Delta V}{V}$ which depends only on the three normal strains.

2.3 Mohr's circle for simple load cases
Draw MOHR's circle for a uniaxial tensile and compression test (see Fig. 2.1), a pure shear test (see Fig. 2.4), and a hydrostatic compression test (see Fig. 2.6).

2.4 Derivation of the evaluation equation for the torsion test
Derive the equation to evaluate the shear modulus G from the pure shear test as shown in Fig. 2.4a, i.e. $G = \frac{M_{\mathrm{T}} L}{I_{\mathrm{p}} \vartheta_{\mathrm{L}}}$.

2.5 Alternative realization of pure shear test
Sketch MOHR's circle for pure shear test (torsional loading of a cylindrical specimen) and derive a strategy for a pure shear test which is only based on normal stresses.

2.6 Evaluation of isostatic compression test
Simplify the generalized HOOKE's law in terms of shear and bulk modulus to evaluate an isostatic compression test, i.e. to derive the bulk modulus based on $K = \frac{\sigma_{\mathrm{m}}}{\varepsilon_{\mathrm{v}}} = \frac{1/3\sigma_{kk}}{\varepsilon_{kk}}$.

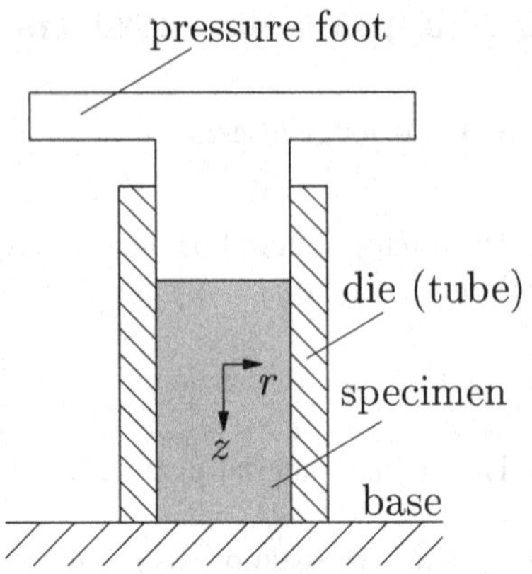

Fig. 2.13 Schematic representation of the axial compaction experiment

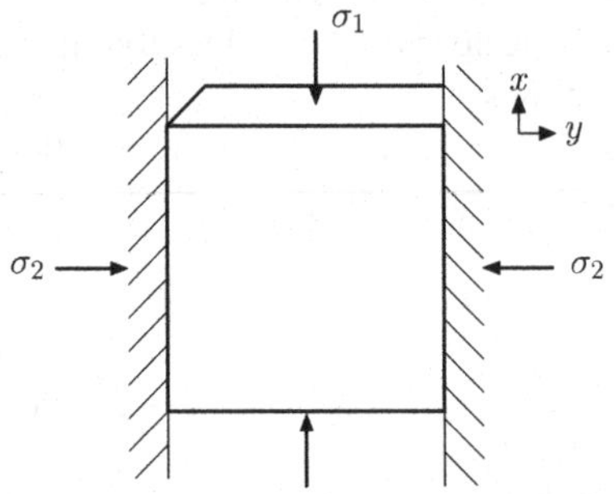

Fig. 2.14 Schematic representation of the plane strain experiment

2.7 Axial compaction

Consider the axial compaction as schematically shown in Fig. 2.13. Derive under the assumption that the wall friction can be neglected an equation for the gradient $d\sigma_z/d\varepsilon_z$.

2.8 Plane strain experiment

Consider the plane strain experiment as schematically shown in Fig. 2.14. A specimen is compressed in the x-direction while the deformation in the y-direction is constrained to zero ($\varepsilon_y = 0$). The z-direction is free, i.e. $\varepsilon_z \neq 0$. Derive from the generalized HOOKE's law a simple equation which allows to calculate POISSON's ratio based on the stress fraction σ_1 and σ_2 where σ_2 is the stress required to prevent any deformation in the y-direction.

2.9 Hooke's law in terms of Lamé's constants

Use a computer algebra system (e.g. Maple®, Mathematica® or Matlab®) to derive the generalized HOOKE's law for a linear isotropic material in terms of LAMÉ'S constants μ and λ in the elastic stiffness form ($\sigma = \sigma(\varepsilon)$) and the elastic compliance form ($\varepsilon = \varepsilon(\sigma)$).

Chapter 3
Elasto-Plastic Material Behavior

Abstract This chapter introduces first the basic equations of plasticity theory. The yield condition, the flow rule, and the hardening rule are introduced. After deriving and presenting these equations for the one-dimensional stress and strain state, the equations are generalized for a three-dimensional state in the scope of the von Mises and Tresca yield condition. The chapter closes with classical failure and fracture hypotheses without the consideration of damage effects.

3.1 One-Dimensional Theory

Let us look again on the engineering stress-strain diagram of a ductile aluminum alloy as shown in Fig. 3.1. Before necking, the stress state is uniaxial and from a practical point of view, damage effects can be disregarded. Only after necking, a three-dimensional stress state is acting (see Fig. 3.2) which is covered in Sect. 3.2.

The characteristic feature of plastic material behavior is that a remaining strain $\varepsilon^{\mathrm{pl}}$ occurs after complete unloading, see Fig. 3.3a. Only the elastic strains $\varepsilon^{\mathrm{el}}$ returns to zero at complete unloading. An additive composition of the strains by their elastic and plastic parts

$$\varepsilon = \varepsilon^{\mathrm{el}} + \varepsilon^{\mathrm{pl}} \tag{3.1}$$

is permitted when restricted to small strains. The elastic strains $\varepsilon^{\mathrm{el}}$ can hereby be determined via HOOKE's law, whereby ε in Eq. (2.3) has to be substituted by $\varepsilon^{\mathrm{el}}$.

Furthermore, no explicit correlation is given anymore for plastic material behavior in general between stress and strain, since the strain state is also dependent on the loading history. Due to this, rate equations are necessary and need to be integrated throughout the entire load history. Within the framework of the time-independent plasticity investigated here, the rate equations can be simplified to incremental relations. From Eq. (3.1) the additive composition of the strain increments results in:

$$\mathrm{d}\varepsilon = \mathrm{d}\varepsilon^{\mathrm{el}} + \mathrm{d}\varepsilon^{\mathrm{pl}} . \tag{3.2}$$

© Springer Science+Business Media Singapore 2016
A. Öchsner, *Continuum Damage and Fracture Mechanics*,
DOI 10.1007/978-981-287-865-6_3

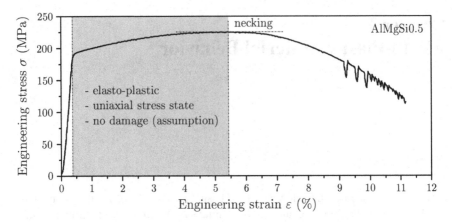

Fig. 3.1 Engineering stress-strain diagram highlighting the region of uniaxial stress during elasto-plastic deformation

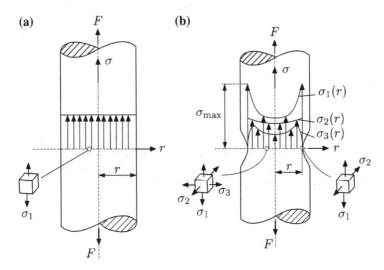

Fig. 3.2 Stress distribution in a round tensile specimen: **a** before and **b** after necking

The constitutive description of plastic material behavior includes

- a yield condition,
- a flow rule and
- a hardening law.

In the following, the case of the monotonic loading is considered first, so that isotropic hardening is explained first in the case of material hardening. This important case, for example, occurs in experimental mechanics at the uniaxial tensile test with monotonic loading. Furthermore, it is assumed that the yield stress is identical in the tensile and compressive regime: $k_t = k_c = k$.

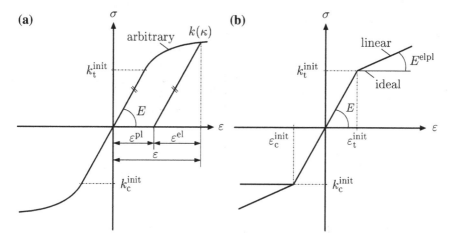

Fig. 3.3 Uniaxial stress-strain diagrams for different isotropic hardening laws: **a** arbitrary hardening; **b** linear hardening and ideal plasticity

3.1.1 Yield Condition

The yield condition enables one to determine whether the relevant material suffers only elastic or also plastic strains at a certain stress state at a point of the relevant body. In the uniaxial tensile test, plastic flow begins when reaching the initial yield stress k^{init}, see Fig. 3.3. The yield condition in its general one-dimensional form can be set as follows ($\mathbb{R} \times \mathbb{R} \to \mathbb{R}$):

$$F = F(\sigma, \kappa), \qquad (3.3)$$

where κ represents the inner variable of isotropic hardening. In the case of ideal plasticity, see Fig. 3.3b, the following is valid: $F = F(\sigma)$. The values of F have the following mechanical meaning, see Fig. 3.4:

$$F(\sigma, \kappa) < 0 \quad \to \quad \text{elastic material behavior}, \qquad (3.4)$$
$$F(\sigma, \kappa) = 0 \quad \to \quad \text{plastic material behavior}, \qquad (3.5)$$
$$F(\sigma, \kappa) > 0 \quad \to \quad \text{invalid}. \qquad (3.6)$$

Fig. 3.4 Schematic representation of the values of the yield condition and the direction of the stress gradient in the uniaxial stress space

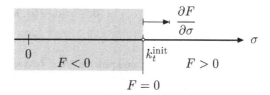

Fig. 3.5 Flow curve for
different isotropic hardening
laws

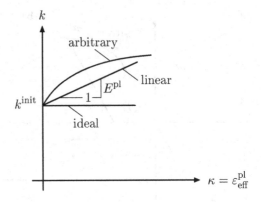

A further simplification results under the assumption that the yield condition can
be split into a pure stress fraction $f(\sigma)$, the so-called yield criterion,[1] and into an
experimental material parameter $k(\kappa)$, the so-called flow stress:

$$F(\sigma, \kappa) = f(\sigma) - k(\kappa). \tag{3.7}$$

For a uniaxial tensile test (see Fig. 3.3) the yield condition can be noted in the
following form:

$$F(\sigma, \kappa) = |\sigma| - k(\kappa) \leq 0. \tag{3.8}$$

If one considers the idealized case of linear hardening (see Fig. 3.3b), Eq. (3.8) can
be written as

$$F(\sigma, \kappa) = |\sigma| - (k^{\text{init}} + E^{\text{pl}}\kappa) \leq 0. \tag{3.9}$$

The parameter E^{pl} is the plastic modulus (see Fig. 3.5), which becomes zero in the
case of ideal plasticity:

$$F(\sigma, \kappa) = |\sigma| - k^{\text{init}} \leq 0. \tag{3.10}$$

3.1.2 Flow Rule

The flow rule serves as a mathematical description of the evolution of the infinitesimal
increments of the plastic strain $d\varepsilon^{\text{pl}}$ in the course of the load history of the body. In
its most general one-dimensional form, the flow rule can be set up as follows [59]:

$$d\varepsilon^{\text{pl}} = d\lambda\, r(\sigma, \kappa), \tag{3.11}$$

---

[1]If the unit of the yield criterion equals the stress, $f(\sigma)$ represents the equivalent stress or effective
stress. In the general three-dimensional case the following is valid under consideration of the
symmetry of the stress tensor $\sigma_{\text{eff}} : (\mathbb{R}^6 \rightarrow \mathbb{R}_+)$.

whereupon the factor $d\lambda$ is described as the consistency parameter ($d\lambda \geq 0$) and $r : (\mathbb{R} \times \mathbb{R} \to \mathbb{R})$ as the function of the flow direction.[2] One considers that solely for $d\varepsilon^{pl} = 0$ then $d\lambda = 0$ results. Based on the stability postulate of DRUCKER[3] [18] the following flow rule can be derived[4]:

$$d\varepsilon^{pl} = d\lambda \frac{\partial F(\sigma, \kappa)}{\partial \sigma}. \qquad (3.12)$$

Such a flow rule is referred to as the normal rule[5] (see Fig. 3.4a) or due to $r = \partial F(\sigma, \kappa)/\partial \sigma$ as the *associated* flow rule.

Experimental results, among other things from the area of granular materials [10] can however be approximated better if the stress gradient is substituted through a different function, the so-called plastic potential Q. The resulting flow rule is then referred to as the *non-associated* flow rule:

$$d\varepsilon^{pl} = d\lambda \frac{\partial Q(\sigma, \kappa)}{\partial \sigma}. \qquad (3.13)$$

In the case of quite complicated yield conditions, often it occurs that a simpler yield condition is used for Q in the first approximation, for which the gradient can easily be determined.

The application of the associated flow rule (3.12) to the yield conditions according to Eqs. (3.8)–(3.10) yields for all three types of yield conditions (meaning arbitrary hardening, linear hardening and ideal plasticity):

$$d\varepsilon^{pl} = d\lambda \operatorname{sgn}(\sigma), \qquad (3.14)$$

where $\operatorname{sgn}(\sigma)$ represents the so-called sign function,[6] which can adopt the following values:

$$\operatorname{sgn}(\sigma) = \begin{cases} -1 & \text{for } \sigma < 0 \\ 0 & \text{for } \sigma = 0 \\ +1 & \text{for } \sigma > 0 \end{cases}. \qquad (3.15)$$

---

[2] In the general three-dimensional case r hereby defines the direction of the vector $d\varepsilon^{pl}$, while the scalar factor defines the absolute value.

[3] Daniel Charles DRUCKER (1918–2001), US engineer.

[4] A formal alternative derivation of the associated flow rule can occur via the LAGRANGE multiplier method as extreme value with side-conditions from the principle of maximum plastic work [8].

[5] In the general three-dimensional case the image vector of the plastic strain increment has to be positioned upright and outside oriented to the yield surface, see Fig. 3.4b.

[6] Also signum function; from the Latin 'signum' for 'sign'.

3.1.3 Hardening Rule

The hardening law allows the consideration of the influence of material hardening on the yield condition and the flow rule.

3.1.3.1 Isotropic Hardening

In the case of isotropic hardening, the yield stress is expressed as being dependent on an inner variable κ:

$$k = k(\kappa).\qquad(3.16)$$

If the equivalent plastic strain[7] is used for the hardening variable ($\kappa = |\varepsilon^{\text{pl}}|$), then one talks about strain hardening.

Another possibility is to describe the hardening being dependent on the specific[8] plastic work ($\kappa = w^{\text{pl}} = \int \sigma \mathrm{d}\varepsilon^{\text{pl}}$). Then one talks about work hardening. If Eq. (3.16) is combined with the flow rule according to (3.14), the evolution equation for the isotropic hardening variable results in:

$$\mathrm{d}\kappa = \mathrm{d}|\varepsilon^{\text{pl}}| = \mathrm{d}\lambda.\qquad(3.17)$$

Figure 3.5 shows the flow curve, meaning the graphical illustration of the yield stress being dependent on the inner variable for different hardening approaches.

The yield condition which was expressed in Eq. (3.3) can be generalized to the formulation

$$F = F(\sigma, q) = 0,\qquad(3.18)$$

where the internal variable q considers the influence of the material hardening on the yield condition. The evolution equation for this internal variable can be stated in its most general form based on Eq. (3.18) as

$$\mathrm{d}q = \mathrm{d}\lambda \times h(\sigma, q),\qquad(3.19)$$

---

[7] The effective plastic strain is in the general three-dimensional case the function $\varepsilon^{\text{pl}}_{\text{eff}} : (\mathbb{R}^6 \to \mathbb{R}_+)$. In the one-dimensional case, the following is valid: $\varepsilon^{\text{pl}}_{\text{eff}} = \sqrt{\varepsilon^{\text{pl}}\varepsilon^{\text{pl}}} = |\varepsilon^{\text{pl}}|$. Attention: Finite element programs optionally use the more general definition for the illustration in the post processor, this means $\varepsilon^{\text{pl}}_{\text{eff}} = \sqrt{\frac{2}{3} \sum \Delta\varepsilon^{\text{pl}}_{ij} \sum \Delta\varepsilon^{\text{pl}}_{ij}}$, which considers the lateral contraction at uniaxial stress problems in the plastic area via the factor $\frac{2}{3}$. However in pure one-dimensional problems *without* lateral contraction, this formula leads to an illustration of the effective plastic strain, which is reduced by the factor $\sqrt{\frac{2}{3}} \approx 0.816$.

[8] This is the volume-specific definition, meaning $\left[w^{\text{pl}}\right] = \frac{\text{N}}{\text{m}^2}\frac{\text{m}}{\text{m}} = \frac{\text{kg m}}{\text{s}^2\text{m}^2}\frac{\text{m}}{\text{m}} = \frac{\text{kg m}^2}{\text{s}^2\text{m}^3} = \frac{\text{J}}{\text{m}^3}$.

where the function h defines the evolution of the hardening parameter. Assigning for the internal variable $q = \kappa$ (in the case that κ equals the effective plastic strain, one speaks of a strain space formulation) and considering the case of associated plasticity, a more specific rule for the evolution of the internal variable is given as

$$\mathrm{d}\kappa = -\mathrm{d}\lambda \times \left(D^{\mathrm{pl}}\right)^{-1} \frac{\partial F(\sigma, \kappa)}{\partial \kappa} = -\mathrm{d}\lambda \times \frac{\partial \kappa}{\partial k(\kappa)} \frac{\partial F(\sigma, \kappa)}{\partial \kappa}$$

$$= -\mathrm{d}\lambda \times \frac{1}{E^{\mathrm{pl}}} \frac{\partial F(\sigma, \kappa)}{\partial \kappa}, \tag{3.20}$$

where D^{pl} is the generalized plastic modulus. Considering the yield stress k as the internal variable, one obtains a stress space formulation as $F = F(\sigma, k)$ and the corresponding evolution equation for the internal variable is given by:

$$\mathrm{d}k = -\mathrm{d}\lambda \times D^{\mathrm{pl}} \frac{\partial F(\sigma, k)}{\partial k} = -\mathrm{d}\lambda \times E^{\mathrm{pl}} \frac{\partial F(\sigma, k)}{\partial k}, \tag{3.21}$$

where $\mathrm{d}k$ can be written as $E^{\mathrm{pl}}\mathrm{d}\kappa$. Thus, one may alternatively formulate:

$$\mathrm{d}\kappa = -\mathrm{d}\lambda \times \frac{\partial F(\sigma, k)}{\partial k}. \tag{3.22}$$

Application of the instruction for the evolution of the internal variable according to Eqs. (3.20) or (3.22) to the yield condition with $k_{\mathrm{t}} = k_{\mathrm{t}}(\kappa)$ (cf. Eqs. (3.9) and (3.7)) gives:

$$\mathrm{d}\kappa = \mathrm{d}\lambda. \tag{3.23}$$

Thus, it turned out that h, i.e. the evolution equation for the hardening parameter in Eq. (3.19), is simplified to $h = 1$. However, in the case of more complex yield conditions, the function h may take a more complex form. This will be shown in Chap. 4 where the GURSON's damage model is introduced.

3.1.3.2 Kinematic Hardening

In the case of pure monotonic loading, i.e. pure tensile or pure compression, it is not possible to distinguish the cases of isotropic or kinematic hardening from the stress-strain diagram. Let us look in the following at a uniaxial test with plastic deformation and stress reversal as schematically shown in Fig. 3.6. The test starts without any initial stress or strain in the origin of the stress-strain diagram (point '0') and a tensile load is continuously increased. The first part of the path, i.e. as long as the stress is below the yield stress k, is in the pure linear-elastic range and HOOKE's law describes the stress-strain behavior. Reaching the yield stress k (point

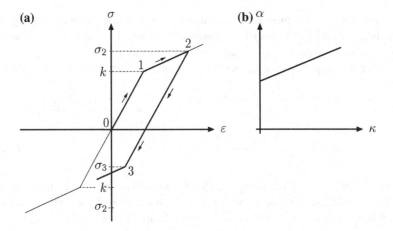

Fig. 3.6 Uniaxial kinematic hardening: **a** idealized stress-strain curve with BAUSCHINGER effect and **b** kinematic hardening parameter as a function of the internal variable (linear hardening)

'1'), the slope of the diagram changes and plastic deformation occurs. With ongoing increasing load, the plastic deformation and the plastic strain increases in this part of the diagram. Let us assume now that the load is reversed at point '2'. The unloading is completely elastic and compressive stress develops as soon as the load path passes the strain axis. The interesting question is now when the subsequent plastic deformations starts in the compressive regime. This plastic deformation occurs now in the case of kinematic hardening at a stress level σ_3 which is lower than the initial yield stress k or the subsequent stress σ_2. This behavior is known as the BAUSCHINGER[9] effect [6] and requires plastic pre-straining with subsequent load reversal.

The behavior shown in Fig. 3.6a can be described based on the following yield condition

$$F = |\sigma - \alpha(\kappa)| - k = 0,\tag{3.24}$$

where the initial yield stress k is constant and the kinematic hardening parameter[10] α is a function of an internal variable κ. Figure 3.6b shows the case of linear hardening where a linear relationship between kinematic hardening parameter and internal variable is obtained.

The simplest relation between the kinematic hardening parameter and the internal variable was proposed in [35] as

$$\alpha = H\varepsilon^{\text{pl}} \quad \text{or} \quad d\alpha = H d\varepsilon^{\text{pl}},\tag{3.25}$$

---

[9]Johann BAUSCHINGER (1834–1893), German mathematician and engineer.

[10]An alternative expression for the kinematic hardening parameter is back-stress.

where H is a constant called the kinematic hardening modulus and the plastic strain is assigned as the internal variable. Thus, Eq. (3.25) describes the case of linear hardening. A more general formulation of Eq. (3.25) is known as PRAGER's[11] hardening rule [47, 48]:

$$d\alpha = H(\sigma, \kappa_i)d\varepsilon^{pl}, \qquad (3.26)$$

where the kinematic hardening modulus is now a scalar function which depends on the state variables (σ, κ_i). One suggestion is to use the effective plastic strain ε^{pl}_{eff} as the internal variable [5]. A further extension is proposed in [29] where the hardening modulus is formulated as a tensor.

Another formulation was proposed by ZIEGLER[12] [57, 71] as

$$d\alpha = d\mu(\sigma - \alpha), \qquad (3.27)$$

where the proportionality factor $d\mu$ can be expressed as:

$$d\mu = ad\varepsilon^{pl}, \qquad (3.28)$$

or in a more general way as $a = a(\sigma, \kappa_i)$. The rule given in Eq. (3.27) is known in the literature as ZIEGLER's hardening rule. It should be noted here that the plastic strain increments in Eqs. (3.28) and (3.26) can be calculated based on the flow rules given in Sect. 3.1.2. Thus, the kinematic hardening rules can be expressed in a more general way as:

$$d\alpha = d\lambda h(\sigma, \alpha). \qquad (3.29)$$

3.1.3.3 Combined Hardening

The isotropic and kinematic hardening rules presented in Sects. 3.1.3.1 and 3.1.3.2 can simply be joined together to obtain a combined hardening rule for the one-dimensional yield condition as:

$$F(\sigma, \mathbf{q}) = |\sigma - \alpha| - k(\kappa), \qquad (3.30)$$

or for the special case of isotropic linear hardening as

$$F = |\sigma - \alpha| - (k^{init} + E^{pl}\varepsilon^{pl}_{eff}), \qquad (3.31)$$

---

[11] William PRAGER (1903–1980), German engineer and applied mathematician.
[12] Hans ZIEGLER (1910–1985), Swiss scientist.

where the back-stress α can be a function as indicated in Eqs. (3.25)–(3.27). The associated flow rule is then obtained according to Eq. (3.12) as:

$$d\varepsilon^{pl} = d\lambda \frac{\partial F}{\partial \sigma} = d\lambda \, \text{sgn}(\sigma - \alpha) \tag{3.32}$$

and the isotropic and kinematic hardening (PRAGER) laws can be written as:

$$d\kappa = d|\varepsilon^{pl}| = |d\lambda \, \text{sgn}(\sigma - \alpha)| = d\lambda , \tag{3.33}$$

$$d\alpha = H d\varepsilon^{pl} = H \, d\lambda \, \text{sgn}(\sigma - \alpha) . \tag{3.34}$$

The last two equations can be combined and generally expressed as:

$$dq = d\lambda h(\sigma, q) , \tag{3.35}$$

or

$$\begin{bmatrix} d\kappa \\ d\alpha \end{bmatrix} = d\lambda \begin{bmatrix} 1 \\ H \, \text{sgn}(\sigma - \alpha) \end{bmatrix} . \tag{3.36}$$

3.1.4 Elasto-Plastic Modulus

The stiffness of a material changes during plastic deformation and the strain state is dependent on the loading history. Therefore, HOOKE's law, which is valid for the linear-elastic material behavior according to Eq. (2.3), must be replaced by the following infinitesimal incremental relation:

$$d\sigma = E^{elpl} d\varepsilon , \tag{3.37}$$

where E^{elpl} is the elasto-plastic modulus. The algebraic expression for this modulus can be obtained in the following manner. The total differential of a yield condition $F = F(\sigma, q)$, see Eq. (3.30), is given by:

$$dF(\sigma, q) = \left(\frac{\partial F}{\partial \sigma} \right) d\sigma + \left(\frac{\partial F}{\partial q} \right)^{T} dq = 0 . \tag{3.38}$$

If HOOKE's law (2.3) and the flow rule (3.11) are introduced in the relation for the additive composition of the elastic and plastic strain according to Eq. (3.2), one obtains:

$$d\varepsilon = \frac{1}{E} d\sigma + d\lambda r . \tag{3.39}$$

Multiplication of Eq. (3.39) from the left-hand side with $\left(\frac{\partial F}{\partial \sigma}\right) E$ and inserting in Eq. (3.38) gives, under the consideration of the evolution equation of the hardening variables (3.35), the consistence parameter as:

$$d\lambda = \frac{\left(\frac{\partial F}{\partial \sigma}\right) E}{\left(\frac{\partial F}{\partial \sigma}\right) Er - \left(\frac{\partial F}{\partial q}\right)^{\mathrm{T}} h} d\varepsilon . \tag{3.40}$$

This equation for the consistency parameter can be inserted into Eq. (3.39) and solving for $\frac{d\sigma}{d\varepsilon}$ gives the elasto-plastic modulus as:

$$E^{\text{elpl}} = E - \frac{E \left(\frac{\partial F}{\partial \sigma}\right) Er}{\left(\frac{\partial F}{\partial \sigma}\right) Er - \left(\frac{\partial F}{\partial q}\right)^{\mathrm{T}} h} . \tag{3.41}$$

Let us consider now the case of combined linear kinematic and isotropic hardening (see Eq. (3.31)) where the kinematic hardening modulus H (PRAGER) and the plastic modulus E^{pl} are constant. Furthermore, the flow rule is assumed to be associated. We assume in the following that the yield condition is a function of the following internal variables: $F = F(\sigma, q) = F(\sigma, \varepsilon^{\text{pl}}, \varepsilon^{\text{pl}}_{\text{eff}})$. The corresponding terms in the expression for the elasto-plastic modulus are as follows:

- $\left(\frac{\partial F}{\partial \sigma}\right) = \text{sgn} \left|\sigma - H\varepsilon^{\text{pl}}\right| ,$ \hfill (3.42)

- $r = \frac{\partial F}{\partial \sigma} = \text{sgn} \left|\sigma - H\varepsilon^{\text{pl}}\right| ,$ \hfill (3.43)

- $\left(\frac{\partial F}{\partial q}\right) = \left[\begin{array}{c} -E^{\text{pl}} \\ -\text{sgn} \left(\sigma - H\varepsilon^{\text{pl}}\right) H \end{array} \right] ,$ \hfill (3.44)

- $h = \left[\begin{array}{c} 1 \\ \text{sgn} \left(\sigma - H\varepsilon^{\text{pl}}\right) \end{array} \right] .$ \hfill (3.45)

Introducing these four expressions in Eq. (3.41) finally gives:

$$E^{\text{elpl}} = \frac{d\sigma}{d\varepsilon} = \frac{E(H + E^{\text{pl}})}{E + (H + E^{\text{pl}})} . \tag{3.46}$$

A slightly different derivation is obtained by considering the yield condition depending on the following internal variables: $F = F(\sigma, q) = F(\sigma, \alpha, \varepsilon^{\text{pl}}_{\text{eff}})$. Then, the following different expressions are obtained:

Table 3.1 Comparison of the different definitions of the stress-strain characteristics (moduli) in the case of the one-dimensional $\sigma - \varepsilon$ space

Range	Definition	Graphical representation
Elastic	$E = \dfrac{\mathrm{d}\sigma}{\mathrm{d}\varepsilon^{\mathrm{el}}}$	Figure 3.3
Plastic	$E^{\mathrm{elpl}} = \dfrac{\mathrm{d}\sigma}{\mathrm{d}\varepsilon}$ for $\varepsilon > \varepsilon^{\mathrm{init}}$	Figure 3.3b
	$E^{\mathrm{pl}} = \dfrac{\mathrm{d}k}{\mathrm{d}|\varepsilon^{\mathrm{pl}}|}$	Figure 3.5

$$\bullet \ \left(\frac{\partial F}{\partial q}\right) = \begin{bmatrix} -E^{\mathrm{pl}} \\ -\mathrm{sgn}\left(\sigma - H\varepsilon^{\mathrm{pl}}\right) \end{bmatrix}, \tag{3.47}$$

$$\bullet \ h = \begin{bmatrix} 1 \\ H\,\mathrm{sgn}\left(\sigma - H\varepsilon^{\mathrm{pl}}\right) \end{bmatrix}, \tag{3.48}$$

which result again in Eq. (3.46). The different general definitions of the moduli used in this derivation are summarized in Table 3.1.

At the end of this section, Table 3.2 compares the different equations and formulations of one-dimensional plasticity with the general three-dimensional representations (see for example [7, 59]).

Table 3.2 Comparison between 1D plasticity with combined hardening (E^{pl} and H are assumed constant) and general 3D plasticity

1D *linear* hardening plasticity	General 3D plasticity
Yield condition	
$F = (\sigma, q) \leq 0$	$F(\sigma, q) \leq 0$
$F = |\sigma - \alpha| - (k^{\mathrm{init}} + E^{\mathrm{pl}}\varepsilon_{\mathrm{eff}}^{\mathrm{pl}}) \leq 0$	
Flow rule	
$\mathrm{d}\varepsilon^{\mathrm{pl}} = \mathrm{d}\lambda \times r(\sigma, q)$	$\mathrm{d}''^{\mathrm{pl}} = \mathrm{d}\lambda \times r(\sigma, q)$
$\mathrm{d}\varepsilon^{\mathrm{pl}} = \mathrm{d}\lambda \times \mathrm{sgn}(\sigma - \alpha)$	
Hardening law	
$q = [\mathrm{d}\varepsilon_{\mathrm{eff}}^{\mathrm{pl}}, \alpha]^{\mathrm{T}}$	$q = [\kappa, \alpha]^{\mathrm{T}}$
$\mathrm{d}q = \mathrm{d}\lambda \times h(\sigma, q)$	$\mathrm{d}q = \mathrm{d}\lambda \times h(\sigma, q)$
$\mathrm{d}\varepsilon_{\mathrm{eff}}^{\mathrm{pl}} = \mathrm{d}\lambda, \ \mathrm{d}\alpha = \mathrm{d}\lambda H\,\mathrm{sgn}(\sigma - \alpha)$	
Elasto-plastic modulus/matrix	
$E^{\mathrm{elpl}} = \dfrac{E \times (H + E^{\mathrm{pl}})}{E + (H + E^{\mathrm{pl}})}$	$C^{\mathrm{elpl}} = \left(C - \dfrac{\left(C\frac{\partial F}{\partial \sigma}\right) \times (Cr)^{\mathrm{T}}}{\left(\frac{\partial F}{\partial \sigma}\right)^{\mathrm{T}} Cr - \left(\frac{\partial F}{\partial q}\right)^{\mathrm{T}} h} \right)$

3.1.5 Consideration of Unloading, Reversed Loading and Cyclic Loading

The previous sections considered only monotonic loading either in the tensile or compressive regimes. We will now briefly look at the cases where the loading direction can change. Figure 3.7a shows the case of loading in the elastic ($0 \rightarrow 1$) and elasto-plastic ($1 \rightarrow 2$) range, followed by elastic unloading ($2 \rightarrow 3$) and elastic reloading ($3 \rightarrow 2$).

In the case of Fig. 3.7b, the elastic unloading ($2 \rightarrow 3$) is followed by reversed loading ($3 \rightarrow 4$). The important feature which should be highlighted here is that the unloading phase ($2 \rightarrow 3$) can be described based on HOOKE's law, cf. Eq. (2.3). Figure 3.8 shows the case of cyclic loading where a specimen is exposed to fluctuating loads $F(t)$.

Some characteristic stress quantities are indicated in Fig. 3.8b: The stress range $\Delta\sigma$ is the difference between the maximum and minimum stress:

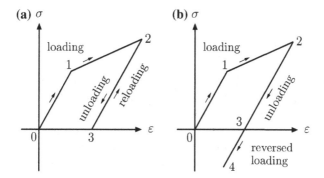

Fig. 3.7 Idealized stress-strain curve with **a** loading—unloading—reloading and **b** loading—unloading—reversed loading

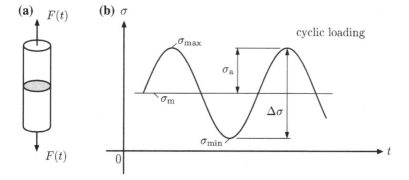

Fig. 3.8 Cyclic loading: **a** idealized specimen and **b** stress-time curve at constant amplitude

$$\Delta \sigma = \sigma_{\max} - \sigma_{\min}. \tag{3.49}$$

The stress amplitude σ_a is half the value of the stress range:

$$\sigma_a = \frac{\Delta \sigma}{2} = \frac{\sigma_{\max} - \sigma_{\min}}{2}. \tag{3.50}$$

The so-called stress ratio R is often used to characterize the stress level in cyclic tests:

$$R = \frac{\sigma_{\min}}{\sigma_{\max}}, \tag{3.51}$$

where a value $R = -1$ characterizes a fully-reversed load cycle, $R = 1$ stands for static loading and $R = 0$ refers to the case where the mean stress is positive and equal to the stress amplitude. In materials testing, cyclic tests are performed to determine the fatigue life of components and structures. Further details can be found in [46, 56].

3.2 Three-Dimensional Behavior

Let us again have a look on a typical engineering stress-strain diagram for a ductile metal as shown in Fig. 3.9. After the necking phenomenon, the stress state in the grey shaded region is multiaxial (see also Fig. 3.2). For the following derivations, we assume that there are no voids, pores (damage) or microcracks (fracture) in the material present. This is a very strong simplification at larger strains and only Chaps. 4 and 5 will treat these effects.

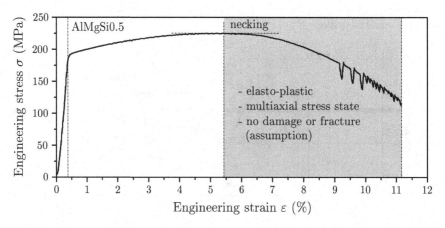

Fig. 3.9 Engineering stress-strain diagram highlighting the region of multiaxial stress during elasto-plastic deformation

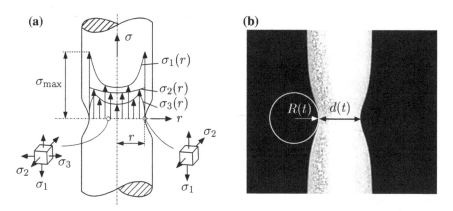

Fig. 3.10 **a** Stress distribution in a round tensile sample after necking; **b** optical determination of minimum cross section d and radius of curvature R. Adapted from [42]

Let us have here a closer look on the stress state in the specimen after necking, see Fig. 3.10a. As already mentioned in Sect. 3.1, three stress components are acting and the sole calculation of the uniaxial true stress does not sufficiently characterize the stress state.

An approximation to convert this multiaxial stress state (σ_1, σ_2, σ_3) in an effective or equivalent stress σ_{eff} (scalar value) was proposed by BRIDGMAN[13] [11, 23, 27, 63] and is based on a correction of the uniaxial true stress σ_{tr} due to the actual geometry of the neck region, i.e. the radius of curvature R and the diameter d (see Fig. 3.10b):

$$\sigma_{\text{eff}} = \frac{\sigma_{\text{tr}}}{\left(1 + \frac{4R}{d}\right) \times \ln\left(1 + \frac{d}{4R}\right)}. \tag{3.52}$$

It should be noted here that the actual geometry (radius of curvature R and the diameter d) can be obtained from image analyses as indicated in Fig. 3.10b.

Not only the stress state is multiaxial after necking but also the strain state. After necking, the flow curve is given in engineering practice[14] as $\sigma_{\text{eff}} = \sigma_{\text{eff}}(\varepsilon_{\text{eff}}^{\text{pl}})$, i.e. as a function of the equivalent plastic strain. Neglecting the elastic strain and assuming that the volume remains constant, the following approximation of the true plastic strain can be derived [16]:

$$\varepsilon_{\text{tr}}^{\text{pl}} = 2 \times \ln\left(\frac{d_0}{d}\right), \tag{3.53}$$

where d is the actual (see Fig. 3.10b) and d_0 the initial diameter of the round tensile sample. Equation (3.53) can be easily derived from Eq. (1.4), i.e. $\varepsilon_{\text{tr}} = \ln(1 + \varepsilon)$.

---

[13]Percy Williams BRIDGMAN (1882–1961), American physicist.

[14]This formulation would be expected in a finite element code.

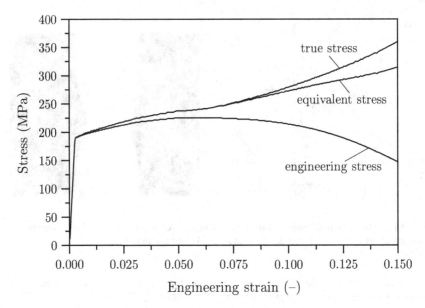

Fig. 3.11 Typical representation of the engineering, true and equivalent stress for a ductile material

Assuming that the elastic strain can be disregarded, one obtains that $\varepsilon_{tr}^{pl} \approx \varepsilon_{tr}$. Under the additional assumption that the material is incompressible (i.e. no volume change: $A_0 L_0 = AL \Leftrightarrow \frac{L}{L_0} = \frac{A_0}{A}$), one can write that:

$$\varepsilon_{tr}^{pl} = \ln\left(1 + \frac{\Delta L}{L_0}\right) = \ln\left(\frac{L}{L_0}\right) = \ln\left(\frac{A_0}{A}\right) = \ln\left(\frac{\frac{\pi}{4}d_0^2}{\frac{\pi}{4}d^2}\right) = \ln\left(\frac{d_0}{d}\right)^2 . \quad (3.54)$$

It should be noted here that this true plastic strain is often equated to the equivalent plastic strain in engineering practice: $\varepsilon_{tr}^{pl} \approx \varepsilon_{eff}^{pl}$. A typical representation of the different stress formulations is shown in Fig. 3.11 where it can be seen that the equivalent stress is situated between the engineering and true stress formulation.

The importance to consider a multiaxial strain state is highlighted in Fig. 3.12 where it can be seen that Eq. (3.53) gives a significant different result for the strain compared to the uniaxial approaches.

3.2.1 Comments on the Stress Tensor

Let us consider again a three-dimensional body which is sufficiently supported and loaded as shown in Fig. 2.9. The stress components acting on a differential volume element may have, for example, the values as shown in Eq. (3.55) for the given

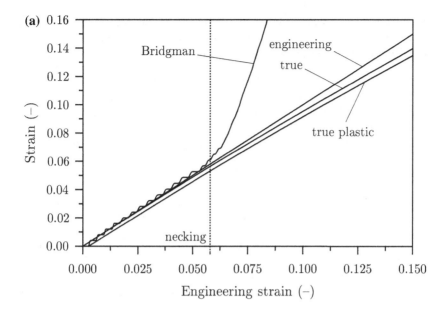

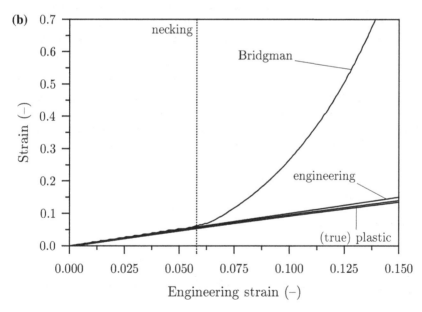

Fig. 3.12 Comparison of different approaches to calculate the strain after necking: **a** smaller strain values and **b** larger strain values

(x, y, z) coordinate system. A transformation from the (x, y, z) to the (x', y', z') coordinate system results in a stress tensor with different stress components, while a principal axis transformation (PAT) calculates the principal stresses σ_i, $(i = 1, 2, 3)$.

$$\sigma_{ij} = \begin{bmatrix} 50 & 0 & 20 \\ 0 & 80 & 20 \\ 20 & 20 & 90 \end{bmatrix}_{(x,y,z)} \overset{\text{rotation}}{\Rightarrow} \begin{bmatrix} 65 & 15 & 28.28427 \\ 15 & 65 & 0 \\ 28.28427 & 0 & 90 \end{bmatrix}_{(x',y',z')}$$

$$\overset{\text{PAT}}{\Rightarrow} \begin{bmatrix} 110 & 0 & 0 \\ 0 & 70 & 0 \\ 0 & 0 & 40 \end{bmatrix}_{(1,2,3)} . \tag{3.55}$$

Looking at this example, the following characteristics can be stated:

- The components of the stress tensor depend on the orientation of the coordinate system.
- There is a specific coordinate system $(1, 2, 3)$ where the shear stresses vanish and only normal stresses remain on the main diagonal, i.e. the so-called principal stresses σ_i $(i = 1, 2, 3)$.
- The six or three stress components cannot easily be compared to experimental values from uniaxial tests (e.g. the initial yield stress in tension $k_{\text{t}}^{\text{init}}$).
- A graphical representation of any surface is much easier in a principal stress space $(1, 2, 3)$ with three coordinates than in a (x, y, z) space with six coordinates.

Let us review in the following the determination of the principal stresses and the axes directions of the corresponding $(1, 2, 3)$ coordinate system. From a mathematical point of view, this question can be answered by determining the eigenvalues of the stress tensor (principal stresses) and the corresponding eigenvectors (principal directions). The solution of the so-called characteristic equation, i.e.

$$\det \left(\sigma_{ij} - \sigma_i \mathbf{I} \right) = 0 , \tag{3.56}$$

gives the three principal stresses σ_i $(i = 1, 2, 3)$. Equation (3.56) can be written in components as:

$$\det \left(\begin{bmatrix} \sigma_{xx} & \sigma_{xy} & \sigma_{xz} \\ \sigma_{yx} & \sigma_{yy} & \sigma_{yz} \\ \sigma_{zx} & \sigma_{zy} & \sigma_{zz} \end{bmatrix} - \sigma_i \begin{bmatrix} 1 & 0 & 0 \\ 0 & 1 & 0 \\ 0 & 0 & 1 \end{bmatrix} \right) = \det \left(\begin{bmatrix} \sigma_{xx} - \sigma_i & \sigma_{xy} & \sigma_{xz} \\ \sigma_{yx} & \sigma_{yy} - \sigma_i & \sigma_{yz} \\ \sigma_{zx} & \sigma_{zy} & \sigma_{zz} - \sigma_i \end{bmatrix} \right) = 0 . \tag{3.57}$$

The calculation of the determinant (see Eq. (A.27)) results in the following cubic equation in σ_i:

$$\sigma_i^3 - \underbrace{\left(\sigma_{xx} + \sigma_{yy} + \sigma_{zz}\right)}_{I_1} \sigma_i^2$$

$$+ \underbrace{\left(\sigma_{xx}\sigma_{yy} + \sigma_{xx}\sigma_{zz} + \sigma_{yy}\sigma_{zz} - \sigma_{xy}^2 - \sigma_{xz}^2 - \sigma_{yz}^2\right)}_{I_2} \sigma_i$$

$$- \underbrace{\left(\sigma_{xx}\sigma_{yy}\sigma_{zz} - \sigma_{xx}\sigma_{yz}^2 - \sigma_{yy}\sigma_{xz}^2 - \sigma_{zz}\sigma_{xy}^2 + 2\sigma_{xy}\sigma_{xz}\sigma_{yz}\right)}_{I_3} = 0 , \qquad (3.58)$$

or in short:

$$\sigma_i^3 - I_1\sigma_i^2 + I_2\sigma_i - I_3 = 0 , \qquad (3.59)$$

where the three roots $(\sigma_1, \sigma_2, \sigma_3)$ of Eq. (3.59) are the principal stresses. Equation (3.58) can be used to define the three scalar so-called principal invariants I_1, I_2 and I_3. These tensor invariants are independent of the orientation of the coordinate system (objectivity) and represent the physical content of the stress tensor.

The coordinates of the ith eigenvector (x_i, y_i, z_i)—which correspond to the direction of one of the new $(1, 2, 3)$ coordinate axes—result from the following system of three equations:

$$\begin{bmatrix} \sigma_{xx} - \sigma_i & \sigma_{xy} & \sigma_{xz} \\ \sigma_{yx} & \sigma_{yy} - \sigma_i & \sigma_{yz} \\ \sigma_{zx} & \sigma_{zy} & \sigma_{zz} - \sigma_i \end{bmatrix} \begin{bmatrix} x_i \\ y_i \\ z_i \end{bmatrix} = \begin{bmatrix} 0 \\ 0 \\ 0 \end{bmatrix} . \qquad (3.60)$$

Let us mention at this point that the determination of the eingenvalues and eigenvectors is also common in applied mechanics for other tensors or matrices. The second moment of the area tensor (frequently also named as 'moment of inertia') has similar properties as the stress tensor:

$$\begin{bmatrix} I_{xx} & I_{xy} & I_{xz} \\ I_{yx} & I_{yy} & I_{yz} \\ I_{zx} & I_{zy} & I_{zz} \end{bmatrix}_{(x,y,z)} \overset{\text{PAT}}{\Rightarrow} \begin{bmatrix} I_1 & 0 & 0 \\ 0 & I_2 & 0 \\ 0 & 0 & I_3 \end{bmatrix}_{(1,2,3)} . \qquad (3.61)$$

The application to the principal equation for a dynamic system results in the eigenfrequencies and eigenmodes of a vibrating system:

$$M\ddot{u} + Ku = 0 \quad \Rightarrow \quad \det(K - \omega_i^2 M) = 0, \qquad (3.62)$$

where the ω_i are the eigenfrequencies.

$$(K - \omega_i^2 M)\Phi = 0 ; \qquad (3.63)$$

where the Φ_i are the eigenmodes of the system.

Let us look in the following a bit closer at the stress invariants.[15] Another interpretation of the principal stress invariants is given by:

- $I_1 = $ trace[16] of σ_{ij}:

$$I_1 = \sigma_{xx} + \sigma_{yy} + \sigma_{zz}. \tag{3.64}$$

- $I_2 = $ sum of the two-row main subdeterminants:

$$I_2 = \begin{vmatrix} \sigma_{xx} & \sigma_{xy} \\ \sigma_{xy} & \sigma_{yy} \end{vmatrix} + \begin{vmatrix} \sigma_{yy} & \sigma_{yz} \\ \sigma_{yz} & \sigma_{zz} \end{vmatrix} + \begin{vmatrix} \sigma_{xx} & \sigma_{xz} \\ \sigma_{xz} & \sigma_{zz} \end{vmatrix}. \tag{3.65}$$

- $I_3 = $ determinant of σ_{ij}:

$$I_3 = \begin{vmatrix} \sigma_{xx} & \sigma_{xy} & \sigma_{xz} \\ \sigma_{xy} & \sigma_{yy} & \sigma_{yz} \\ \sigma_{xz} & \sigma_{yz} & \sigma_{zz} \end{vmatrix}. \tag{3.66}$$

Besides these principal invariants, there is also often another set of invariants used. This set is included in the principal invariants and called basic invariants:

$$J_1 = I_1, \tag{3.67}$$
$$J_2 = \tfrac{1}{2}I_1^2 - I_2, \tag{3.68}$$
$$J_3 = \tfrac{1}{3}I_1^3 - I_1 I_2 + I_3. \tag{3.69}$$

The definition of both sets of invariants is given in Table 3.3.

Table 3.3 Definition of the principal (I_i) and basic (J_i) stress invariants

First stress invariant of σ_{ij}
$I_1 = \sigma_{xx} + \sigma_{yy} + \sigma_{zz}$
$J_1 = \sigma_{xx} + \sigma_{yy} + \sigma_{zz}$
Second stress invariant of σ_{ij}
$I_2 = \sigma_{xx}\sigma_{yy} + \sigma_{xx}\sigma_{zz} + \sigma_{yy}\sigma_{zz} - \sigma_{xy}^2 - \sigma_{xz}^2 - \sigma_{yz}^2$
$J_2 = \tfrac{1}{2}\left(\sigma_{xx}^2 + \sigma_{yy}^2 + \sigma_{zz}^2\right) + \sigma_{xy}^2 + \sigma_{xz}^2 + \sigma_{yz}^2$
Third stress invariant of σ_{ij}
$I_3 = \sigma_{xx}\sigma_{yy}\sigma_{zz} - \sigma_{xx}\sigma_{yz}^2 - \sigma_{yy}\sigma_{xz}^2 - \sigma_{zz}\sigma_{xy}^2 + 2\sigma_{xy}\sigma_{xz}\sigma_{yz}$
$J_3 =$
$\tfrac{1}{3}\left(\sigma_{xx}^3 + \sigma_{yy}^3 + \sigma_{zz}^3 + 3\sigma_{xy}^2\sigma_{xx} + 3\sigma_{xy}^2\sigma_{yy} + 3\sigma_{xz}^2\sigma_{xx} + 3\sigma_{xz}^2\sigma_{zz} + 3\sigma_{yz}^2\sigma_{yy} + 3\sigma_{yz}^2\sigma_{zz} + 6\sigma_{xy}\sigma_{xz}\sigma_{yz}\right)$

---

[15]It is useful for some applications (e.g. the calculation of derivative with respect to the stresses) to *not* consider the symmetry of the shear stress components and to work with nine stress components. These invariants are denoted by $\underline{I}_i$ and $\underline{J}_i$.

[16]The trace of a tensor is the sum of the diagonal elements.

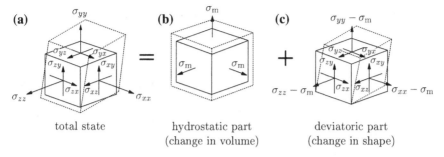

Fig. 3.13 Decomposition of the stress tensor (a) into its spherical (b) and the deviatoric (c) part

It is common in the framework of the plasticity theory of isotropic materials to decompose the stress tensor σ_{ij} into a pure volume changing (spherical or hydrostatic) tensor σ_{ij}^o and a pure shape changing (deviatoric) stress tensor s_{ij} (cf. Fig. 3.13)[17]:

$$\sigma_{ij} = \sigma_{ij}^o + s_{ij} = \sigma_m \mathbf{I} + s_{ij}. \tag{3.70}$$

In Eq. (3.70), $\sigma_m = \frac{1}{3}(\sigma_{xx} + \sigma_{yy} + \sigma_{zz})$ denotes the mean normal stress[18] and $\mathbf{I}$ the identity matrix. Furthermore, EINSTEIN's[19] summation convention was used [36].

Equation (3.70) can be written in components as

$$\underbrace{\begin{bmatrix} \sigma_{xx} & \sigma_{xy} & \sigma_{xz} \\ \sigma_{xy} & \sigma_{yy} & \sigma_{yz} \\ \sigma_{xz} & \sigma_{yz} & \sigma_{zz} \end{bmatrix}}_{\text{stress tensor } \sigma_{ij}} = \underbrace{\begin{bmatrix} \sigma_m & 0 & 0 \\ 0 & \sigma_m & 0 \\ 0 & 0 & \sigma_m \end{bmatrix}}_{\text{hydrostatic tensor } \sigma_{ij}^o} + \underbrace{\begin{bmatrix} s_{xx} & s_{xy} & s_{xz} \\ s_{xy} & s_{yy} & s_{yz} \\ s_{xz} & s_{yz} & s_{zz} \end{bmatrix}}_{\text{deviatoric tensor } s_{ij}}. \tag{3.71}$$

It can be seen that the elements outside the main diagonal, i.e. the shear stresses, are the same for the stress and the deviatoric stress tensor

$$s_{ij} = \sigma_{ij} \quad \text{for} \quad i \neq j, \tag{3.72}$$

$$s_{ij} = \sigma_{ij} - \sigma_m \quad \text{for} \quad i = j, \tag{3.73}$$

and it can be shown that the so-called deviator equation

$$s_{xx} + s_{yy} + s_{zz} = 0 \tag{3.74}$$

---

[17] It should be noted that in the case of anisotropic materials, a hydrostatic stress state may result in a shape change, [8].

[18] Also called the hydrostatic stress; in the context of soil mechanics, the pressure $p = -\sigma_m$ is also used.

[19] Albert EINSTEIN (1879–1955), German theoretical physicist.

holds. The following list summarises the calculation of the stress deviator components:

$$S_{xx} = \sigma_{xx} - \sigma_{\mathrm{m}} = \frac{2}{3}\sigma_{xx} - \frac{1}{3}(\sigma_{yy} + \sigma_{zz}), \tag{3.75}$$

$$S_{yy} = \sigma_{yy} - \sigma_{\mathrm{m}} = \frac{2}{3}\sigma_{yy} - \frac{1}{3}(\sigma_{xx} + \sigma_{zz}), \tag{3.76}$$

$$S_{zz} = \sigma_{zz} - \sigma_{\mathrm{m}} = \frac{2}{3}\sigma_{zz} - \frac{1}{3}(\sigma_{xx} + \sigma_{yy}), \tag{3.77}$$

$$S_{xy} = \sigma_{xy}, \tag{3.78}$$

$$S_{yz} = \sigma_{yz}, \tag{3.79}$$

$$S_{xz} = \sigma_{xz}. \tag{3.80}$$

The hydrostatic part of σ_{ij} has in the case of metallic materials (full dense materials) for temperatures approximately under $0.3T_{\mathrm{kf}}$ (T_{kf}: melting temperature) nearly no influence on the occurrence of inelastic strains since dislocations slip only under the influence of shear stresses (for higher temperatures from 0.3 till $0.5T_{\mathrm{kf}}$ also non-conservative climbing is possible) [61]. On the other hand, the hydrostatic stress has a considerable influence on the yielding behavior in the case of soil mechanics, cellular materials or in damage mechanics (formation of pores, e. g. [26]).

The definition of the invariants summarized in Table 3.3 can be also applied to the split of the stress tensor as outlined in Eq. (3.71), i.e. to evaluate the three invariants for the stress tensor as well as for the hydrostatic and deviatoric tensor. This evaluation is presented in Table 3.4 for the basic invariants J_i. It can be seen in Table 3.4 that the spherical tensor is completely characterised by its first invariant because the second and third invariant are powers of it. The stress deviator tensor is completely characterised by its second and third invariant. Therefore, the physical contents of the stress state σ_{ij} can be described either by the three basic stress invariants J_i or if we use the decomposition in its spherical and deviatoric part by the first invariant of the spherical tensor and the second and third invariant of the stress deviator tensor. In the following, we will only use these three basic invariants to describe yield and failure conditions. Thus, the physical content of a state of stress will be described by the following set of invariants:

$$\sigma_{ij} \;\rightarrow\; J_1^{\mathrm{o}}, J_2', J_3'. \tag{3.81}$$

3.2.2 Graphical Representation of Yield Conditions

Plastic flow starts in a uniaxial tensile test as soon as the acting stress σ reaches the initial yield stress k^{init}. Under a multiaxial stress state, the comparison is replaced by

Table 3.4 Basic invariants in terms of σ_{ij} and principal values

Invariants	General values	Principal values
Stress tensor		
J_1	$\sigma_{xx} + \sigma_{yy} + \sigma_{zz}$	$\sigma_1 + \sigma_2 + \sigma_3$
J_2	$\frac{1}{2}\left(\sigma_{xx}^2 + \sigma_{yy}^2 + \sigma_{zz}^2\right) + \sigma_{xy}^2 + \sigma_{xz}^2 + \sigma_{yz}^2$	$\frac{1}{2}\left(\sigma_1^2 + \sigma_2^2 + \sigma_3^2\right)$
J_3	$\frac{1}{3}\big(\sigma_{xx}^3 + \sigma_{yy}^3 + \sigma_{zz}^3 + 3\sigma_{xy}^2\sigma_{xx} + 3\sigma_{xy}^2\sigma_{yy} + 3\sigma_{xz}^2\sigma_{xx} + 3\sigma_{xz}^2\sigma_{zz} + 3\sigma_{yz}^2\sigma_{yy} + 3\sigma_{yz}^2\sigma_{zz} + 6\sigma_{xy}\sigma_{xz}\sigma_{yz}\big)$	$\frac{1}{3}\left(\sigma_1^3 + \sigma_2^3 + \sigma_3^3\right)$
Spherical tensor		
$J_1^{\rm o}$	$\sigma_{xx} + \sigma_{yy} + \sigma_{zz}$	$\sigma_1 + \sigma_2 + \sigma_3$
$J_2^{\rm o}$	$\frac{1}{6}\left(\sigma_{xx} + \sigma_{yy} + \sigma_{zz}\right)^2$	$\frac{1}{6}\left(\sigma_1 + \sigma_2 + \sigma_3\right)^2$
$J_3^{\rm o}$	$\frac{1}{9}\left(\sigma_{xx} + \sigma_{yy} + \sigma_{zz}\right)^3$	$\frac{1}{9}\left(\sigma_1 + \sigma_2 + \sigma_3\right)^3$
Stress deviator tensor		
J_1'	0	0
J_2'	$\frac{1}{6}\left[(\sigma_{xx} - \sigma_{yy})^2 + (\sigma_{yy} - \sigma_{zz})^2 + (\sigma_{zz} - \sigma_{xx})^2\right] + \sigma_{xy}^2 + \sigma_{yz}^2 + \sigma_{zx}^2 + (\sigma_3 - \sigma_1)^2]$	$\frac{1}{6}\left[(\sigma_1 - \sigma_2)^2 + (\sigma_2 - \sigma_3)^2\right.$
J_3'	$s_{xx}s_{yy}s_{zz} + 2\sigma_{xy}\sigma_{yz}\sigma_{zx} - s_{xx}\sigma_{yz}^2 - s_{yy}\sigma_{zx}^2 - s_{zz}\sigma_{xy}^2$	$s_1 s_2 s_3$
With	$s_{xx} = \frac{1}{3}(2\sigma_{xx} - \sigma_{yy} - \sigma_{zz})$	$s_1 = \frac{1}{3}(2\sigma_1 - \sigma_2 - \sigma_3)$
	$s_{yy} = \frac{1}{3}(-\sigma_{xx} + 2\sigma_{yy} - \sigma_{zz})$	$s_2 = \frac{1}{3}(-\sigma_1 + 2\sigma_2 - \sigma_3)$
	$s_{zz} = \frac{1}{3}(-\sigma_{xx} - \sigma_{yy} + 2\sigma_{zz})$	$s_3 = \frac{1}{3}(-\sigma_1 - \sigma_2 + 2\sigma_3)$

the yield condition. To this end, a scalar value is calculated from the acting six stress components and compared to an experimental scalar value. The yield condition in stress space can be expressed in its most general form ($\mathbb{R}^6 \times \mathbb{R}^{\dim(q)} \to \mathbb{R}$) as:

$$F = F(\boldsymbol{\sigma}, \boldsymbol{q}) . \qquad (3.82)$$

For further characterisation, we assume in the following the special case of ideal plastic material behavior (vector of hardening variables[20] $\boldsymbol{q} = \boldsymbol{0}$) so that for ($\mathbb{R}^6 \to \mathbb{R}$)

$$F = F(\boldsymbol{\sigma}) \qquad (3.83)$$

depends now only on the stress state. The values of F have—as in the uniaxial case—the following mechanical meaning:

---

[20] See Sect. 3.1.3.3.

$$F(\boldsymbol{\sigma}) = 0 \rightarrow \text{plastic material behavior,} \qquad (3.84)$$

$$F(\boldsymbol{\sigma}) < 0 \rightarrow \text{elastic material behavior,} \qquad (3.85)$$

$$F(\boldsymbol{\sigma}) > 0 \rightarrow \text{invalid.} \qquad (3.86)$$

A further simplification is obtained under the assumption that the yield condition can be split in a pure stress part $f(\boldsymbol{\sigma})$, the so-called yield criterion, and an experimental material parameter k:

$$F(\boldsymbol{\sigma}) = f(\boldsymbol{\sigma}) - k. \qquad (3.87)$$

The yield condition $F = 0$ represents in a n-dimensional space a hypersurface that is also called the yield surface or the yield loci. The number n is equal to the independent stress tensor components. A direct graphical representation of the yield surface is not possible due to its dimensionality. However, a reduction of the dimensionality is possible to achieve if a principle axis transformation (see Eq. (3.56)) is applied to the argument σ_{ij}. The components of the stress tensor reduce to the principal stresses σ_1, σ_2 and σ_3 on the principal diagonal of the stress tensor and the non-diagonal elements are equal to zero. In such a principal stress space, it is possible to graphically represent the yield condition as a three-dimensional surface. This space is also called the HAIGH-WESTERGAARD stress space. A hydrostatic stress state lies in such a principal stress system on the space diagonal (hydrostatic axis). Any plane perpendicular to the hydrostatic axis is called an octahedral plane. The particular octahedral plane passing through the origin is called the deviatoric plane or π-plane [15]. Because $\sigma_1 + \sigma_2 + \sigma_3 = 0$, it follows from Eq. (3.70) that $\sigma_{ij} = s_{ij}$, i.e. any stress state on the π-plane is pure deviatoric.

The possibility of a representation of a yield condition based on a set of independent stress invariants (e.g. according to Eq. (3.81)) is the characteristic of any isotropic yield condition, regardless of the choice of coordinate system. Therefore, Eq. (3.83) can also be written as

$$F = F(J_1^\circ, J_2', J_3'). \qquad (3.88)$$

On the basis of the dependency of the yield condition on the invariants, a descriptive classification can be performed. Yield conditions independent of the hydrostatic stress (J_1°) can be represented by the invariants J_2' and J_3'. Stress states with $J_2' = \text{const.}$ lie on a circle around the hydrostatic axis in an octahedral plane. A dependency of the yield condition on J_3' results in a deviation from the circle shape. The yield surface forms a prismatic body whose longitudinal axis is represented by the hydrostatic axis. A dependency on J_1° denotes a size change of the cross-section of the yield surface along the hydrostatic axis. However, the shape of the cross-section remains similar in the mathematical sense. Therefore, a dependency on J_1° can be represented by sectional views through planes along the hydrostatic axis.

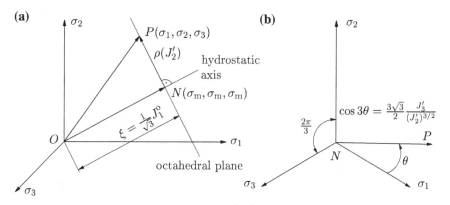

Fig. 3.14 Geometrical interpretation of basic stress invariants: **a** principal stress space; **b** octahedral plane

The geometrical interpretation of stress invariants [15] is given in Fig. 3.14: It can be seen that an arbitrary stress state P can be expressed by its position along the hydrostatic axis $\xi = \frac{1}{\sqrt{3}} J_1^\circ$ and its polar coordinates ($\rho = \sqrt{2J_2'}$, $\theta(J_2', J_3')$) in the octahedral plane through P. For the set of polar coordinates, the so-called stress Lode angle θ is defined in the range $0 \le \theta \le 60°$ as, [39],

$$\cos(3\theta) = \frac{3\sqrt{3}}{2} \cdot \frac{J_3'}{(J_2')^{3/2}}. \tag{3.89}$$

The set of coordinates ($\xi, \rho, \cos(3\theta)$) is known as the HAIGH-WESTERGAARD coordinates. To investigate the shape of the yield surface, multiaxial stress states must be realized and the initial yield points marked and approximated in the HAIGH-WESTERGAARD space.

3.2.3 Mises Yield Condition

The VON MISES[21] yield condition states that plastic deformation starts as soon as the distortional deformation energy per unit volume (see Sect. B.2), i.e.

$$w^s = \frac{1+\nu}{6E} \left[(\sigma_{xx} - \sigma_{yy})^2 + (\sigma_{yy} - \sigma_{zz})^2 + (\sigma_{zz} - \sigma_{xx})^2 + 6(\tau_{xy}^2 + \tau_{yz}^2 + \tau_{xz}^2) \right],$$

$$\tag{3.90}$$

---

[21]Richard Edler VON MISES (1883–1953), Austrian scientist and mathematician.

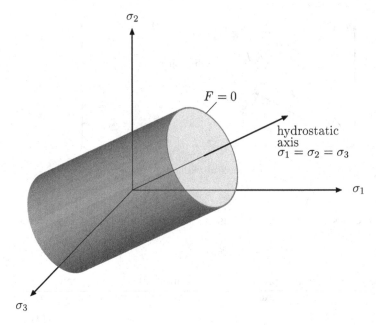

Fig. 3.15 Graphical representation of the yield condition according to VON MISES in the principal stress space

reaches a critical value $(k_t^2/(6G))$ [2, 38]. The expression in units of stress is given for a general three-dimensional stress state as

$$F(\sigma_{ij}) = \underbrace{\sqrt{\frac{1}{2}\left((\sigma_x - \sigma_y)^2 + (\sigma_y - \sigma_z)^2 + (\sigma_z - \sigma_x)^2\right) + 3\left(\sigma_{xy}^2 + \sigma_{yz}^2 + \sigma_{xz}^2\right)}}_{\sigma_{\text{eff}}} - k_t = 0,$$

(3.91)

or expressed with the second invariant of the stress deviator:

$$F(J_2') = \sqrt{3J_2'} - k_t = 0.$$

(3.92)

The graphical representation in the principal stress space is given in Fig. 3.15 where a cylinder with its longitudinal axis equal to the hydrostatic axis is obtained.

The view along the hydrostatic axis is shown in Fig. 3.16a where it can be seen that there is no difference under tension and compression for uniaxial stress states. The representation in the $\sqrt{3J_2'} - J_1^\circ$ space (see Fig. 3.16b) shows that the yield condition is independent of the hydrostatic stress.

Representations in the two-component principal $\sigma_1 - \sigma_2$ and the normal/shear $\sigma - \tau$ space are obtained as, see Fig. 3.17:

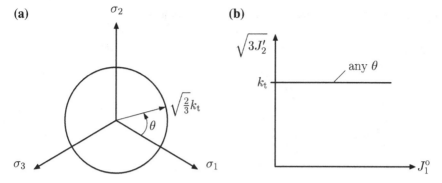

Fig. 3.16 Graphical representation of the yield condition according to VON MISES: **a** octahedral plane; **b** $\sqrt{3J_2'} - J_1^\circ$ space

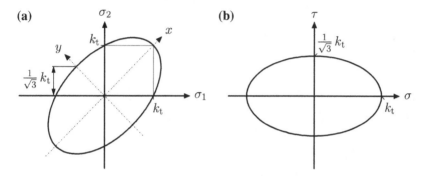

Fig. 3.17 Graphical representation of the yield condition according to VON MISES: **a** $\sigma_1 - \sigma_2$ space; space; **b** $\sigma - \tau$ space

$$F_{\sigma_1 - \sigma_2} = \sigma_1^2 + \sigma_2^2 - \sigma_1\sigma_2 - k_t^2 \,, \tag{3.93}$$

$$F_{\sigma - \tau} = \left(\frac{\sigma}{k_t}\right)^2 + \left(\frac{\sqrt{3}\,\tau}{k_t}\right)^2 - 1 \,. \tag{3.94}$$

Table 3.5 illustrates the fact that it is not the right approach to look on single stress components if one has to judge if the stress state is in the elastic or already in the plastic domain. Only the equivalent stress based on a yield condition can answer this question in the case of multiaxial stress states.

Table 3.5 Equivalent VON MISES stress for different stress states

Stress tensor $\sigma_i j$	von Mises stress Eq. (3.91)	Domain $(k_t^{\text{init}} = 150)$
$\begin{bmatrix} 100 & 0 & 0 \\ 0 & 100 & 0 \\ 0 & 0 & 0 \end{bmatrix}$	100	Elastic
$\begin{bmatrix} 100 & 0 & 0 \\ 0 & -100 & 0 \\ 0 & 0 & 0 \end{bmatrix}$	173.2	Plastic
$\begin{bmatrix} 200 & 0 & 20 \\ 0 & 80 & 20 \\ 20 & 20 & 90 \end{bmatrix}$	125.3	Elastic
$\begin{bmatrix} 200 & 0 & 20 \\ 0 & 80 & 20 \\ 20 & 20 & 200 \end{bmatrix}$	129.3	Elastic
$\begin{bmatrix} 100 & 0 & 20 \\ 0 & 80 & 20 \\ 20 & 20 & -80 \end{bmatrix}$	177.8	Plastic

3.2.4 Tresca Yield Condition

The TRESCA[22] yield condition, also known as the maximum shear stress theory, postulates yielding as soon as the maximum shear stress reaches an experimental value. The expression is given for the principal stresses as

$$\max \left(\frac{1}{2}|\sigma_1 - \sigma_2|, \frac{1}{2}|\sigma_2 - \sigma_3|, \frac{1}{2}|\sigma_3 - \sigma_1| \right) = k_s, \tag{3.95}$$

or

$$F(\sigma_i) = \max \left(\frac{1}{2}|\sigma_1 - \sigma_2|, \frac{1}{2}|\sigma_2 - \sigma_3|, \frac{1}{2}|\sigma_3 - \sigma_1| \right) - k_s = 0. \tag{3.96}$$

Expressed with the second and third invariant of the stress deviator, the following formulation is obtained:

---

[22]Henri Édouard TRESCA, (1814–1885), French mechanical engineer.

$$F(J_2', J_3') = 4\left(J_2'\right)^3 - 27\left(J_3'\right)^2 - 36k_s^2\left(J_2'\right)^2 + 96k_s^4 J_2' - 64k_s^6 = 0. \quad (3.97)$$

The graphical representation in the principal stress space is given in Fig. 3.18 where a prism of six sides with its longitudinal axis equal to the hydrostatic axis is obtained.

The view along the hydrostatic axis is shown in Fig. 3.19a where a hexagon can be seen. The representation in the $\sqrt{3J_2'} - J_1^\circ$ space (see Fig. 3.19b) shows that the yield condition is independent of the hydrostatic stress.

Representations in the two-component principal $\sigma_1 - \sigma_2$ and the normal/shear $\sigma - \tau$ space are obtained as (see Fig. 3.20):

$$
\begin{aligned}
\sigma_2^{(1)} &= \sigma_1 - 2k_s = \sigma_1 - k_t, \\
\sigma_2^{(2)} &= \sigma_1 + 2k_s = \sigma_1 + k_t, \\
\sigma_2^{(3)} &= 2k_s = k_t, \\
\sigma_2^{(4)} &= -2k_s = -k_t, \\
\sigma_1^{(5)} &= 2k_s = k_t, \\
\sigma_1^{(6)} &= -2k_s = -k_t,
\end{aligned}
\quad (3.98)
$$

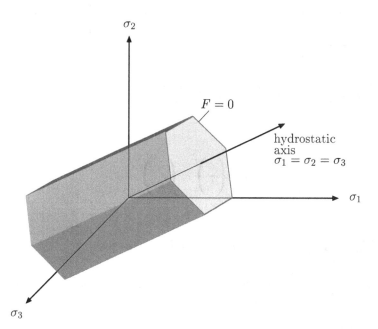

Fig. 3.18 Graphical representation of the yield condition according to TRESCA in the principal stress space

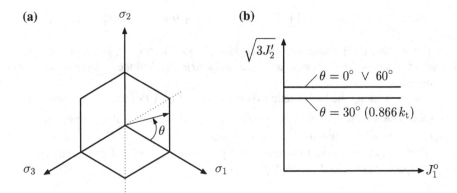

Fig. 3.19 Graphical representation of the yield condition according to TRESCA: **a** octahedral plane; **b** $\sqrt{3J'_2} - J^\circ_1$ space

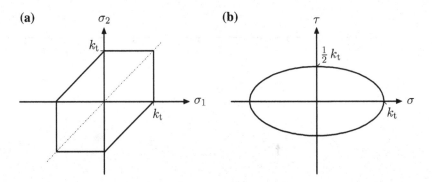

Fig. 3.20 Graphical representation of the yield condition according to TRESCA: **a** $\sigma_1 - \sigma_2$ space; space; **b** $\sigma - \tau$ space

or:

$$F_{\sigma-\tau} = \left(\frac{\sigma}{k_t}\right)^2 + \left(\frac{2\tau}{k_t}\right)^2 - 1\,. \tag{3.99}$$

3.3 Classical Failure and Fracture Hypotheses

Reviewing the previous chapters, it is obvious that 'failure' can have diverse meanings or definitions, such as:

- reaching the plastic range,
- damage (pores/voids),

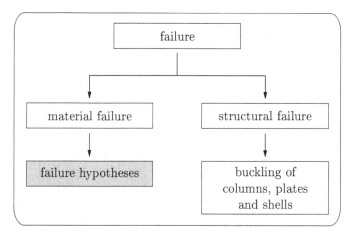

Fig. 3.21 Different theories of failure

- microcracks,
- fracture (separation).

All these forms of failure are known as material failure. Nevertheless, it should be noted here that there is another type of failure which is known as structural failure [65], see Fig. 3.21. The following explanations will be restricted to material failure since the focus of this book is on materials and constitutive equations.

Let us review again the general concept of continuum mechanical modelling, see Fig. 3.22. We already explained this concept in Fig. 2.8 and illustrated the process to derive a partial differential equation which describes the mechanical problem. A solution of the differential equation, for example based on analytical or numerical methods, results in the displacement field of the problem.

Application of the kinematics relation gives the strain field which can be converted based on the constitutive equation into the stress field. Failure and fracture hypotheses take the information of the stress or strain state and predict the failure of a material. Depending on the fundamental behavior of a material or its microstructure, different classical criteria were proposed. A common distinction of materials is based on the stress-strain curve. If the material shows a distinct plastic region (see Fig. 3.23a), it is said to have ductile behavior while a sudden failure in or after the elastic region defines a brittle material (see Fig. 3.23b).

In the case of ductile materials, it might be important to distinguish between classical metals and porous or cellular structures which are sensitive to the hydrostatic pressure in the plastic regime.

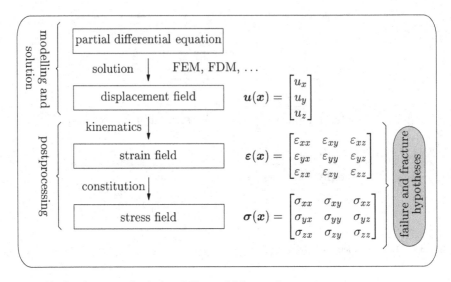

Fig. 3.22 Continuum mechanical modelling and failure analyses

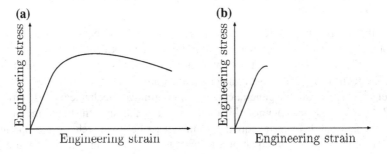

Fig. 3.23 Engineering stress-strain diagrams: **a** ductile material and **b** brittle material

3.3.1 Principal Stress Hypothesis

The principal stress hypothesis[23] was proposed by RANKINE,[24] LAMÉ[25] and NAVIER,[26] and predicts failure as soon as the maximum principal stess reaches a critical value k_t or k_c. The determination of the principal stresses requires the solution of the following equation

---

[23] Alternatively named as the maximum principal stress hypothesis.

[24] William John Macquorn RANKINE (1820–1872), Scottish mechanical engineer, civil engineer, physicist and mathematician.

[25] Gabriel Léon Jean Baptiste LAMÉ (1795–1870), French mathematician.

[26] Claude-Louis NAVIER, (1785–1836), French engineer and physicist.

Fig. 3.24 Graphical representation of the principal stress hypothesis in the principal stress space

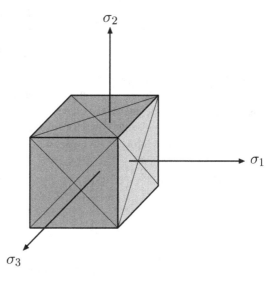

$$\det\left(\begin{bmatrix}\sigma_{xx} & \sigma_{xy} & \sigma_{xz}\\ \sigma_{yx} & \sigma_{yy} & \sigma_{yz}\\ \sigma_{zx} & \sigma_{zy} & \sigma_{zz}\end{bmatrix} - \sigma_i\begin{bmatrix}1 & 0 & 0\\ 0 & 1 & 0\\ 0 & 0 & 1\end{bmatrix}\right) = 0, \tag{3.100}$$

where the calculation of the determinant (see Eq. (A.27)) results in a cubic equation in σ_i. The three roots ($\sigma_1, \sigma_2, \sigma_3$) of this cubic equation are then the principal stresses. Thus, the failure hypothesis can be expressed as follows:

$$\sigma_1 = k_t \vee k_c, \ \sigma_2 = k_t \vee k_c, \ \sigma_3 = k_t \vee k_c. \tag{3.101}$$

This failure surface is a cube in the principal stress space, see Fig. 3.24.
This hypothesis assumes brittle material behavior until failure.

3.3.2 Principal Strain Hypothesis

The principal strain hypothesis[27] was proposed by SAINT-VENANT[28] and BACH,[29] and predicts failure as soon as the maximum principal strain reaches a critical value ε_t. The determination of the principal strains requires the solution of the following equation

---

[27] Alternatively named as the maximum principal strain hypothesis.

[28] Adhémar Jean Claude Barré de SAINT-VENANT (1797–1886), French mechanician and mathematician.

[29] Carl Julius von BACH (1847–1931), German mechanical engineer.

$$\det\left(\begin{bmatrix} \varepsilon_{xx} & \varepsilon_{xy} & \varepsilon_{xz} \\ \varepsilon_{yx} & \varepsilon_{yy} & \varepsilon_{yz} \\ \varepsilon_{zx} & \varepsilon_{zy} & \varepsilon_{zz} \end{bmatrix} - \varepsilon_i \begin{bmatrix} 1 & 0 & 0 \\ 0 & 1 & 0 \\ 0 & 0 & 1 \end{bmatrix}\right) = 0, \tag{3.102}$$

where the calculation of the determinant (see Eq. (A.27)) results in a cubic equation in ε_i. The three roots (ε_1, ε_2, ε_3) of this cubic equation are then the principal strains. Assuming linear-elastic material behavior until failure, the critical stress can be expressed as $k_t = E\varepsilon_t$. Using now HOOKE's law according to Eq. (2.16) to express the principal strains in terms of stresses, i.e.

$$\varepsilon_1 = \frac{1}{E}(\sigma_1 - \nu(\sigma_2 + \sigma_3)), \tag{3.103}$$

$$\varepsilon_2 = \frac{1}{E}(\sigma_2 - \nu(\sigma_1 + \sigma_3)), \tag{3.104}$$

$$\varepsilon_3 = \frac{1}{E}(\sigma_3 - \nu(\sigma_1 + \sigma_2)), \tag{3.105}$$

one can write the failure hypothesis in the principal stress space as follows:

$$\sigma_1 - \nu(\sigma_2 + \sigma_3) = k_t, \tag{3.106}$$
$$\sigma_2 - \nu(\sigma_1 + \sigma_3) = k_t, \tag{3.107}$$
$$\sigma_3 - \nu(\sigma_1 + \sigma_2) = k_t. \tag{3.108}$$

This failure surface is a pyramid with a three-sided base and its side faces around the hydrostatic axis. The apex is given for $\sigma_1 = \sigma_2 = \sigma_3 = \frac{\sigma_t}{1-2\nu}$ (Fig. 3.25).

3.3.3 Strain Energy Hypothesis

The strain energy hypothesis[30] was proposed by BELTRAMI[31] and predicts failure as soon as the total deformation energy per unit volume, i.e.

$$w = \frac{1 - 2\nu}{6E}(\sigma_{xx} + \sigma_{yy} + \sigma_{zz})^2 +$$

$$+ \frac{1 + \nu}{6E}\left[(\sigma_{xx} - \sigma_{yy})^2 + (\sigma_{yy} - \sigma_{zz})^2 + (\sigma_{zz} - \sigma_{xx})^2 + 6(\tau_{xy}^2 + \tau_{yz}^2 + \tau_{xz}^2)\right], \tag{3.109}$$

---

[30] Alternatively named as the maximum strain energy hypothesis.
[31] Eugenio BELTRAMI (1835–1899), Italian mathematician.

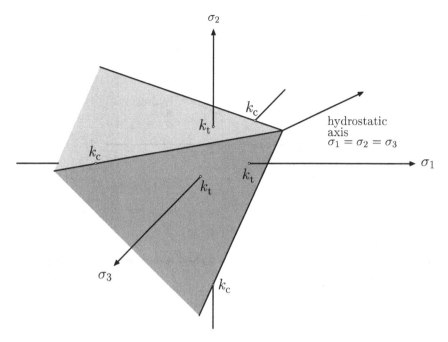

Fig. 3.25 Graphical representation of the principal strain hypothesis in the principal stress space

reaches a critical value w_c. This critical value may be defined as $w_c = \frac{\sigma_c^2}{2E}$ [25]. The graphical representation in the principal stress space is given in Fig. 3.26 as an ellipsoid with major axis equal to the hydrostatic axis. The apex on the hydrostatic axis is equal to:

$$\sigma_1 = \sigma_2 = \sigma_3 = \frac{\pm\,\sigma_c}{\sqrt{3(1-\nu)}} = \pm\sqrt{\frac{2w_c E}{3(1-2\nu)}} = \pm\sqrt{2w_c K}\,. \tag{3.110}$$

This hypothesis assumes linear-elastic material behavior until failure.

3.3.4 Drucker-Prager Hypothesis

The DRUCKER-PRAGER hypothesis can be expressed in terms of invariants as

$$F(J_1^\circ, J_2') = \alpha J_1^\circ + \sqrt{J_2'} - k_s\,, \tag{3.111}$$

where α and k_s are material parameters. For $\alpha = 0$, this hypothesis reduces to the VON MISES condition, see Sect. 3.2.3. The graphical representation in the principal stress

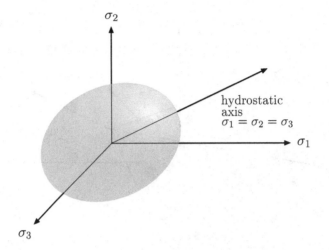

Fig. 3.26 Graphical representation of the strain energy hypothesis in the principal stress space

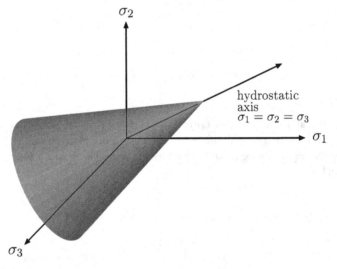

Fig. 3.27 Graphical representation of the yield condition according to DRUCKER-PRAGER in the principal stress space

space is given in Fig. 3.27 as a cone with longitudinal axis equal to the hydrostatic axis.

This hypothesis is commonly used for granular and geological materials with plastic deformations as well as in modified formulation for concrete, composites, foams/sponges, and other pressure-sensitive materials.

3.4 Supplementary Problems

3.1 Knowledge questions

- Consider a ductile metal in a uniaxial tensile test. Characterize the stress state (a) before and (b) after necking. Characterize the specimen's deformation (a) before and (b) after necking
- Sketch the engineering stress-strain diagram for an elastic-ideally plastic material.
- Sketch the engineering stress-strain diagram for a rigid-ideally plastic material.
- Describe the purpose of the (a) yield condition, (b) flow rule and (c) hardening rule.
- Sketch the graphical representation of the (a) elastic, (b) plastic and (c) elasto-plastic modulus.
- How many principal stresses can be extracted from the three-dimensional stress tensor?
- Characterize the deformation which results from the (a) hydrostatic and (b) deviatoric stress state.
- Describe the graphical representation of the VON MISES yield condition in the principal stress space.
- Given is a principal stress state $\sigma_1(>0)$ and $\sigma_2(>0)$. Give the equation for the maximum shear stress as a function of σ_1 and σ_2.
- Give the relation between the initial tensile and shear yield stress for the VON MISES yield condition.
- Give the relation between the initial tensile and shear yield stress for the TRESCA yield condition.
- Can a material fail in a hydrostatic stress state based on the (a) VON MISES yield condition and (b) the principal stress hypothesis?

3.2 Evaluation of tensile test with linear hardening

Figure 3.28 shows an idealized stress-strain diagram as it could, for example, be derived experimentally from an uniaxial tensile test. The right-hand end is displaced by $u = 8 \times 10^{-3}$ m in total.

Calculate for this problem:

- The YOUNG's modulus E,
- the plastic modulus E^{pl},
- the elasto-plastic modulus E^{elpl},
- the equation of the flow curve $k(\kappa)$,
- the stress-strain relation in the elastic part,
- the stress-strain relation in the elasto-plastic part,
- the plastic strains after unloading from 300 and 600 MPa.

3.3 Investigation of initial yielding for different materials and boundary conditions

Consider a simple rod which is at the left-hand side fixed and at the right-hand side loaded by a force F or a displacement u. Describe the procedure to check if the rod is still in the elastic range or already undergoes plastic deformation (Fig. 3.29).

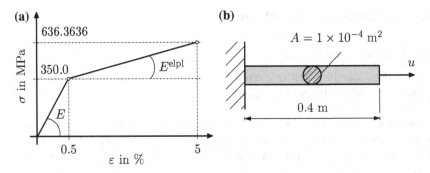

Fig. 3.28 Tensile test: **a** stress-strain diagram and **b** geometry and boundary condition

Fig. 3.29 Simple tensile test with a rod: **a** force boundary condition and **b** displacement boundary condition

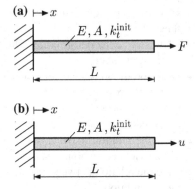

Table 3.6 Common engineering materials

Material	E (MPa)	k_t^{init} (MPa)
Al (7075)	71,000	495
Ti (Ti-6Al-4V)	110,000	910
St (4340)	207,000	1420

Consider the three engineering materials as characterized in Table 3.6.

3.4 Calculation of the principal stresses for a plane stress state

Given is a plane stress state ($\sigma_{xx} = 100, \sigma_{yy} = 80, \sigma_{xy} = 20$). Calculate the principal stresses σ_1 and σ_2 based on:

- the relations derived from MOHR's circle,
- the determination of the eingenvalues of the stress tensor σ_{ij}.

3.5 Calculation of the principal stresses and the corresponding principal directions

Given is a stress tensor in a (x, y, z) coordinate system as shown in Eq. (3.112).

$$\sigma_{ij} = \begin{bmatrix} 50 & 0 & 20 \\ 0 & 80 & 20 \\ 20 & 20 & 90 \end{bmatrix}_{(x,y,z)} \tag{3.112}$$

Calculate the principal stresses and the corresponding principal directions. Show that the obtained unit vectors form a right-handed trihedron.

3.6 Calculation of the principal and basic invariants

Given is a stress tensor in a (x, y, z) coordinate system as shown in Eq. (3.113).

$$\sigma_{ij} = \begin{bmatrix} 50 & 0 & 20 \\ 0 & 80 & 20 \\ 20 & 20 & 90 \end{bmatrix}_{(x,y,z)} \tag{3.113}$$

Calculate the principal and basic invariants and show that the invariants can be converted into each other.

3.7 Difference between shear and tensile yield stress for the von Mises and Tresca yield condition

Consider a two-component $\sigma - \tau$ stress space and calculate the difference between the shear k_s and tensile k_t yield stress for the VON MISES and TRESCA yield conditions.

3.8 Difference between the von Mises and Tresca yield condition in two-component stress spaces

Calculate the maximum difference between the VON MISES and TRESCA yield conditions in the two-component $\sigma - \tau$ and $\sigma_1 - \sigma_2$ stress space.

3.9 Hydrostatic stress and Mises/Tresca yield condition

Given is a hydrostatic stress state $\sigma_1 = \sigma_2 = \sigma_3 = -p$. Apply the VON MISES and TRESCA yield conditions to check if the material reaches its yield stress k.

3.10 Influence of isotropic and kinematic hardening on the shape of the yield surface

Consider the VON MISES yield condition in the $\sigma - \tau$ space. The initial yield surface can be expressed as

$$F_{\sigma-\tau} = \left(\frac{\sigma}{k_t^{init}}\right)^2 + \left(\frac{\sqrt{3}\,\tau}{k_t^{init}}\right)^2 - 1.$$

The corresponding equations for isotropic and kinematic hardening are given as follows:

$$F_{\sigma-\tau} = \left(\frac{\sigma}{k_t(\kappa)}\right)^2 + \left(\frac{\sqrt{3}\,\tau}{k_t(\kappa)}\right)^2 - 1\,,$$

$$F_{\sigma-\tau} = \left(\frac{\sigma - \alpha_\sigma}{k_t^{\text{init}}}\right)^2 + \left(\frac{\sqrt{3}\,(\tau - \alpha_\tau)}{k_t^{\text{init}}}\right)^2 - 1\,.$$

Sketch the initial yield surface and the subsequent yield surfaces for isotropic and kinematic hardening. Conclude which effects the different types of hardening have on the shape of the initial yield surface. Take $k_t(\kappa) = 1.2k_t^{\text{init}}$, $\alpha_\sigma = 0.2k_t^{\text{init}}$ and $\alpha_\tau = 0.1k_t^{\text{init}}$.

3.11 Thin-walled pressure vessel under internal pressure: von Mises and Tresca yield condition

Given is a thin-walled cylindrical pressure vessel (inner diameter $d = 1000$ mm and wall thickness t) as schematically shown in Fig. B.4. The vessel is loaded by an internal pressure $p = 5$ MPa. The initial tensile yield stress is $k_t^{\text{init}} = 320$ MPa. Determine the minimum wall thickness t under consideration of (a) the VON MISES and (b) the TRESCA yield condition to avoid plastic deformation of the vessel.

3.12 Thin-walled pressure vessel under internal pressure and axial force: von Mises and Tresca yield condition

Given is a thin-walled cylindrical pressure vessel (inner diameter $d = 1000$ mm and wall thickness t) as schematically shown in Fig. 3.30. The vessel is loaded by an internal pressure $p = 5$ MPa and an axial force $F = 2200$ kN. The initial tensile yield stress is $k_t^{\text{init}} = 320$ MPa. Determine the minimum wall thickness t under consideration of (a) the VON MISES and (b) the TRESCA yield condition to avoid plastic deformation of the vessel. Disregard all effects from the ends of the cylinder.

3.13 Thin-walled pressure vessel under internal pressure and twisting moment: von Mises and Tresca yield condition

Given is a thin-walled cylindrical pressure vessel (inner diameter $d = 1000$ mm and wall thickness t) as schematically shown in Fig. 3.31. The vessel is loaded by an internal pressure $p = 5$ MPa and a twisting moment $M_T = 10^9$ Nmm. The initial tensile yield stress is $k_t^{\text{init}} = 320$ MPa. Determine the minimum wall thickness t

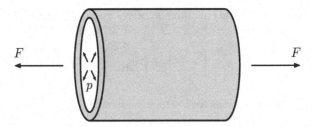

Fig. 3.30 Thin-walled pressure vessel under internal pressure and axial force

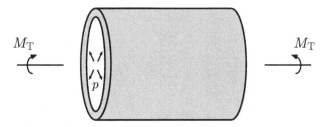

Fig. 3.31 Thin-walled pressure vessel under internal pressure and twisting moment

under consideration of (a) the VON MISES and (b) the TRESCA yield condition to avoid plastic deformation of the vessel. Disregard all effects from the ends of the cylinder.

Chapter 4
Continuum Damage Mechanics

Abstract This chapter introduces the general concept of ductile damage, i.e. the development of microvoids and pores due to large plastic strains and deformations. The general definitions of damage based on a scalar damage variable as the surface damage variable or the void volume fraction are introduced and experimental approaches to measure the quantity of damage are discussed. The chapter introduces two different damage concepts, i.e. the Lemaitre and the Gurson damage model, which are first derived for the one-dimensional case and then briefly generalized for the three-dimensional situation.

4.1 Representation of Damage and Experimental Approaches

Let us have a look again on a typical engineering stress-strain diagram for a ductile metal as shown in Fig. 4.1. As mentioned previously, the stress state after the necking phenomenon is multiaxial (see also Fig. 3.2). For the following derivations, we consider now the existence and development of voids, pores or microcracks in the material. This refined modelling approach is much closer to experimental observations and allows a more accurate description of the material in the elasto-plastic range at larger strains. It should be noted here that the boundaries of this region are not strictly defined. However, it is known from experimental observations that there is a significant increase of damage after the necking of a material [32].

SEM micrographs which correspond to the stress-strain curve shown in Fig. 4.1, i.e. corresponding to the aluminum alloy AlMgSi0.5, were already shown in Fig. 1.3. It could be concluded from these pictures that the formation of pores results from the breaking of the brittle precipitates or the separation from the matrix.

A similar behavior can be observed for the alloy AlCuMg1, see Fig. 4.2 [21]. Energy dispersive X-ray spectroscopy (EDX) analysis revealed that the visible precipitates consists mainly, in addition to aluminum, of silicon and magnesium or copper. This alloy shows again heavy signs of localized damage around the precipitated particles. Pores were found to form mainly at the points of fracture of such brittle particles, and otherwise at locations were the particle has loosened itself from the surrounding matrix.

© Springer Science+Business Media Singapore 2016
A. Öchsner, *Continuum Damage and Fracture Mechanics*,
DOI 10.1007/978-981-287-865-6_4

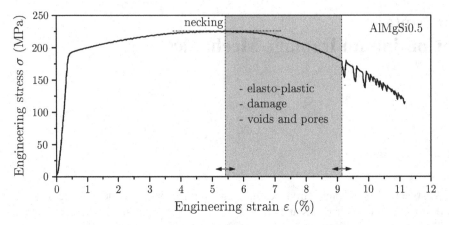

Fig. 4.1 Engineering stress-strain diagram highlighting the region of damage evolution during elasto-plastic deformation

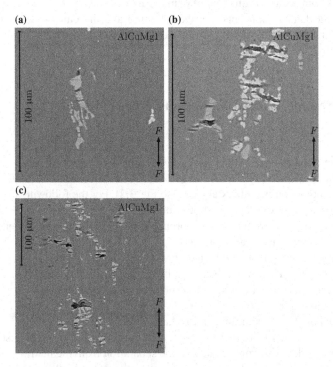

Fig. 4.2 SEM micrographs (longitudinal section, parallel to the loading direction) of the ductile aluminum alloy AlCuMg1: **a** plastic range before maximum stress, **b** plastic range close to maximum stress and **c** close to final failure. Adapted from [21]

(a)   **(b)**

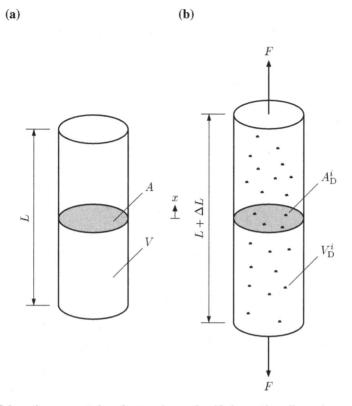

Fig. 4.3 Schematics representation of **a** an undamaged and **b** damaged tensile specimen

Let us consider an idealized uniaxial tensile sample as shown in Fig. 4.3 where the undamaged or initial tensile specimen is shown on the left-hand side and the damaged or deformed specimen on the right-hand side. It should be noted here that the size of the specimen must be in such a range that the considered volume represents a representative volume element (RVE) for the considered material. Some estimates for the minimum size of RVEs for different materials are given in [32] and repeated in Table 4.1.

Let A be the overall cross-sectional area of the specimen (marked in grey in Fig. 4.3a) and A_D be the total area of the micro-cracks and voids in the considered

Table 4.1 Typical sizes of RVEs. Taken from [32]

Material	Size in mm^3
Metals and ceramics	0.1
Polymers and most composites	1
Wood	10
Concrete	100

area which is in Fig. 4.3b marked in black. The effective resisting area is denoted by $\bar{A}$. Based on these quantities, the surface damage variable D can be introduced as:

$$D = \frac{\sum_i A_D^i}{A} = \frac{A_D}{A} = \frac{A - \bar{A}}{A}. \tag{4.1}$$

A state $D = 0$ corresponds to the undamaged state, $D = 1$ represents the rupture of the specimen into two parts and $0 < D < 1$ characterizes the damaged state. Alternatively the damage due to voids in ductile materials is often characterized by the void volume fraction or porosity [25]:

$$f = \frac{\sum_i V_D^i}{V} = \frac{V_D}{V} = \frac{V - \bar{V}}{V}, \tag{4.2}$$

where V_D is the total volume of voids and V is the volume of the RVE.

4.1 Example: Cubic RVE and pore—definition of damage

Let us look in the following on a simple example where a cubic RVE is considered, see Fig. 4.4.

The pore is simplified as a small cube. The initial side length of the undamaged RVE is given as a_i, see Fig. 4.4a. The damaged RVE (see Fig. 4.4b) contains a small cubic pore of side length a_D and the unknown outer side length is denoted by a. Determine based on the assumption of elastic material behavior and that no residual micro-stresses are acting on the damaged RVE the outer diameter a. Then calculate the damage parameter D and the void volume fraction f and relate these results to the relative density $(\varrho - \varrho_i)/\varrho_i$.

(a)   **(b)**

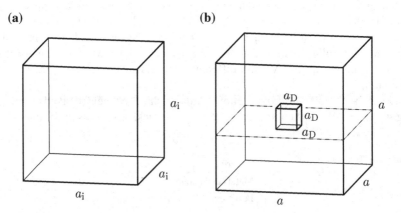

Fig. 4.4 Illustration of damage definition based on a cubic RVE and cubic pore: **a** initial undamaged and **b** damaged state

4.1 Solution

Assuming elastic material behavior and a stress-free configuration, the solid volume in the undamaged and damaged RVE remains the same, i.e.

$$V_i = V - V_D, \tag{4.3}$$

or expressed in terms of the side lengths:

$$a_i^3 = a^3 - a_D^3. \tag{4.4}$$

The last equation can be rearranged for the unknown side length of the damaged RVE:

$$a = \left(a_i^3 + a_D^3\right)^{\frac{1}{3}}. \tag{4.5}$$

To calculate the relative density, it is required to calculate first the densities of the undamaged and the damaged RVE. The undamaged density is obtained as $\varrho_i = \frac{m}{V_i} = \frac{m}{a_i^3}$ while the density of the damaged RVE is obtained with the side length a according to Eq. (4.5) as:

$$\varrho = \frac{m}{a^3} = \frac{m}{a_i^3 + a_D^3}. \tag{4.6}$$

Thus, the relative density for this specific problem can be expressed as:

$$\frac{\varrho - \varrho_i}{\varrho_i} = \frac{\frac{m}{a_i^3 + a_D^3} - \frac{m}{a_i^3}}{\frac{m}{a_i^3}} = \frac{-a_D^3}{a_i^3 + a_D^3}. \tag{4.7}$$

The surface damage variable (see Eq. (4.11)) then reads:

$$D = \frac{A_D}{A} = \frac{a_D^2}{a^2} = \frac{a_D^2}{\left(a_i^3 + a_D^3\right)^{\frac{2}{3}}} = \left(\frac{a_D^3}{a_i^3 + a_D^3}\right)^{\frac{2}{3}} = \left(1 - \frac{\varrho}{\varrho_i}\right)^{\frac{2}{3}}. \tag{4.8}$$

In a similar way, the void volume fraction (see Eq. (4.2)) is obtained as:

$$f = \frac{V_D}{V} = \frac{a_D^3}{a^3} = \frac{a_D^3}{a_i^3 + a_D^3} = 1 - \frac{\varrho}{\varrho_i}. \tag{4.9}$$

Comparing the results from Eqs. (4.8) and (4.9), we can conclude that the evaluation of the damage based on a specific single unit cell results in quite different expressions in terms of D and f.

It must be highlighted here that the simplified case elaborated in Example 4.1 is not representative for irregular microstructures, see Fig. 4.5. It can be shown based

(a)   **(b)**

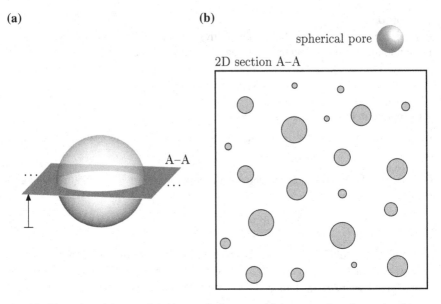

Fig. 4.5 Illustration of damage definition: **a** single pore and **b** larger number of pores in arbitrary arrangement

on stereological methods—applying mathematical relationships between 2D and 3D geometry—that the area and volume fractions must be the same for real microstructures [17, 19, 20, 69]:

$$D = f. \tag{4.10}$$

The experimental measurement of damage can be based on different approaches [32]:

• **Direct measurement of the damaged surface**
Figure 4.6 shows the extraction of a small specimen disk taken at right angles to the loading direction prior to the final failure of the specimen. The damage is then determined by evaluating the damage areas A_D^i on the top and/or bottom surface according to Eq. (4.11).

The summation of the single areas A_D^i can be automatized to a certain extent by converting the 2D images to binary versions and using a digital image analysis software, see Fig. 4.7.

A schematic comparison of this 2D approach, D, with a volumetric measurement (ARCHIMEDES' principle)[1], f, is given in Fig. 4.8. Despite the fact that from a theoretical point of view (see Eq. (4.10)) one should expect equality between these damage parameters, real experimental evaluations may show a slightly different behavior: the two-dimensional parameter may yield a smaller value due to the metallographic

---

[1] ARCHIMEDES of Syracuse (287 BC–ca. 212 BC), Greek mathematician, physicist, engineer, inventor, and astronomer.

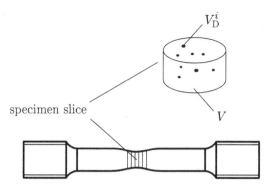

Fig. 4.6 Taking specimen disks from a necked cylindrical specimen. Adapted from [22]

(a)   **(b)**

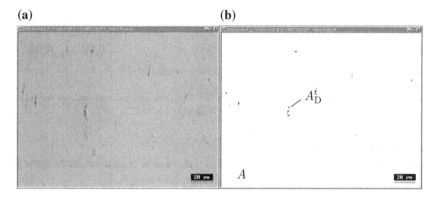

Fig. 4.7 a SEM image of the cross section of a specimen and **b** the binary version in order to better detect the porosity. Adapted from [22]

specimen preparation. During grinding and polishing the metallic matrix deforms plastically causing pores to partially close. This effect can only be avoided by the use of much lengthier and more involved specimen preparation techniques. However, still a linear relationship exists between f and D.

• **Variation of elastic modulus**
The realization of a uniaxial tensile test with reversed loading allows to relate the damage parameter to the YOUNG'S modulus and the modulus of the damaged material, see Sect. 4.2.

• **Ultrasonic waves propagation**
The NEWTON- LAPLACE[2,3] equation, i.e. $c = c(K, \varrho) = c(E, \nu, \varrho)$, allows to measure the damage based on the variation of the speed of sound in an undamaged and damaged specimen disk.

---

[2]Isaac NEWTON (1642–1726/7), English physicist and mathematician.
[3]Pierre-Simon, marquis de LAPLACE (1749–1827), French scientist and mathematician.

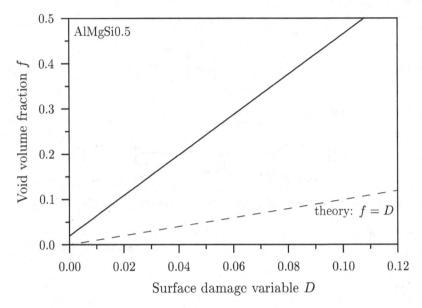

Fig. 4.8 Comparison of the results obtained by 2D digital image analysis and by weighting (ARCHIMEDES' principle) for the parameter f and D. Adapted from [22]

- **Variation of microhardness**
The hardness test based on a small diamond indenter can be used to relate the damage parameter to hardness of the undamaged and damaged material.

- **Variation of density**
The development of pores results in a decrease of the density. A density measurement based on ARCHIMEDES' principle allows to correlate the void volume fraction to the density of the undamaged and damaged material.

- **Variation of electrical resistance**
The definition of the electrical resistance, i.e. $R \sim \frac{L}{A}$, can be used to measure the damage parameter.

- **Direct pore volume measurement**
X-ray computed tomography allows to create accurate three-dimensional models of the specimen and the determination of the void volume fraction.

Let us highlight at the end of this section that the one-dimensional derivations of the following sections are done for an idealized one-dimensional tensile test. The original dimensions of, for example, the cylindrical specimen are characterized by the cross-sectional area A_0 and length L. This specimen is now elongated in a universal testing machine and its length increases to $L + \Delta L$. In the case of a real specimen made of a common engineering material (see Fig. 4.9a), the cross-sectional area would reduce to $A_0 - \Delta A$. This phenomenon could be described based on POISSON's ratio. However, if we assume an idealized uniaxial state, i.e. a

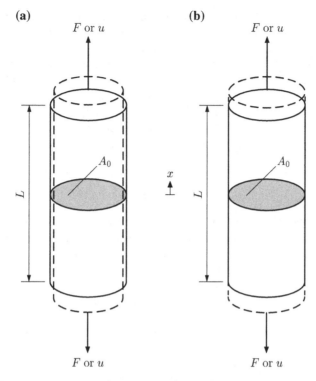

Fig. 4.9 Schematic representation of a uniaxial tensile test based on cylindrical specimens: **a** with contraction and **b** without contraction as a simplified modelling approach

uniaxial stress *and* strain state, the contraction is disregarded and the cross-sectional area is assumed to remain constant, Fig. 4.9b. This simplification allows to derive simpler equations under maintaining of all the required steps as in the more general case.

4.2 Lemaitre's Damage Model

The following section briefly summarizes the major ideas of ductile damage based on the concept given by LEMAITRE in [30, 31]. Let A be the overall cross-sectional area of the specimen (marked in grey in Fig. 4.3a) and A_D be the total area of the micro-cracks and voids in the considered area which is in Fig. 4.3b marked in black. The effective resisting area is denoted by $\bar{A}$. Based on these quantities, the damage variable D can be introduced as:

$$D = \frac{\sum_i A_D^i}{A} = \frac{A_D}{A} = \frac{A - \bar{A}}{A}. \tag{4.11}$$

If this definition is based on a RVE, then the same damage variable is obtained based on the volume of the micro-cracks and voids [69]: $D = V_D/V$. A state $D = 0$ corresponds to the undamaged state, $D = 1$ represents the rupture of the specimen into two parts and $0 < D < 1$ characterizes the damaged state. In the scope of this chapter, an isotropic damage variable is assumed. This means that the defects are equally distributed in all directions of the specimen. Thus, a scalar description of the damage is sufficient under the *hypothesis of isotropy*. If the resisting area in Eq. (4.11) is used to calculate the stress in the specimens, the *concept of effective stress* is obtained which states that the effective stress in the specimen is given by:

$$\bar{\sigma} = \frac{\sigma}{1 - D}. \tag{4.12}$$

It must be stated here that this definition of the effective stress holds only in the tensile regime. Under compression, some defects may close again or in the limiting case, all defects can be closed again so that the effective stress is again equal to the usual stress σ. However, this effect in the compressive regime will be not considered within this chapter. In the case of the strain, the *hypothesis of strain equivalence* is applied which states that the strain behavior of a damaged material is represented by the virgin material:

$$\bar{\varepsilon} = \varepsilon. \tag{4.13}$$

Based on these assumptions and simplifications, HOOKE's law can be written with the effective stress $\bar{\sigma}$ and elastic strain ε^{el} as:

$$\bar{\sigma} = E\varepsilon^{el}, \tag{4.14}$$

which can be expressed with the definition of the effective stress given in Eq. (4.12) as:

$$\sigma = \underbrace{(1 - D)E}_{\bar{E}}\,\varepsilon, \tag{4.15}$$

where E is the elastic modulus of the undamaged material (initial modulus) and $\bar{E}$ is the modulus of the damaged material. The last equation offers an elegant way to experimentally determine the evolution of the damage variable D. Measuring during tests with reversed loading stress and strain based on the usual engineering definitions, i.e. $\sigma = F/A$ and $\varepsilon = \Delta L/L$, the damage variable can be indirectly obtained from the variation of the elasticity modulus as:

$$D = 1 - \frac{\bar{E}}{E}. \tag{4.16}$$

The classical continuum theory of plasticity is based on three equations, i.e. the yield condition, the flow rule and the hardening law. For a one-dimensional stress state, the yield condition reads under consideration of the damage effects as

$$F = \frac{|\sigma|}{1 - D} - k(\kappa), \tag{4.17}$$

where $|\sigma|/(1 - D)$ is the equivalent stress which is compared to the experimental value k. The flow rule and the evolution equation for the internal variable κ do not change and are given by Eqs. (3.12) and (3.17).

In the case of damage mechanics, there is in addition the evolution equation for the damage variable required. Following the notation in [40] and considering a one-dimensional stress state, the model for the ductile damage evolution can be expressed as

$$dD = \frac{d\lambda}{1 - D}\left(\frac{-Y}{r}\right)^s = d|\varepsilon^{pl}|\left(\frac{-Y}{r}\right)^s, \tag{4.18}$$

where Y is the so-called damage energy release rate which corresponds to the variation of internal energy density due to damage growth at constant stress, and r and s are damage evolution material parameters. For a one-dimensional stress state, Y takes the form:

$$Y = -\frac{\sigma^2}{2E(1 - D)^2} \stackrel{4.12}{=} -\frac{\overline{\sigma}^2}{2E}. \tag{4.19}$$

The basic equations for the one-dimensional model according to LEMAITRE are collected in the following Table 4.2.

Table 4.2 Basic equations of the LEMAITRE model (isotropic hardening) in the case of a uniaxial stress state with σ as acting stress

1D LEMAITRE model
HOOKE's Law
$\sigma = (1 - D)E\varepsilon$
Yield condition
$F = \dfrac{|\sigma|}{1 - D} - k(\kappa)$
Flow rule
$d\varepsilon^{pl} = d\lambda \dfrac{\text{sgn}(\sigma)}{1 - D}$
Evolution of hardening variable
$d\kappa = d\lambda$
Evolution of damage variable
$dD = \dfrac{d\lambda}{1 - D}\left(\dfrac{-Y}{r}\right)^s = d|\varepsilon^{pl}|\left(\dfrac{-Y}{r}\right)^s$
with $Y = -\dfrac{\sigma^2}{2E(1 - D)^2}$

An experimental strategy based on a simple tensile test and a fatigue test (WÖHLER curve)[4] for the determination of the parameters r and s is given in [33]. A simpler form of Eq. (4.18) is given in [32] as:

$$dD = d|\varepsilon^{pl}| \times \frac{-Y}{r}, \tag{4.20}$$

or under consideration of Eq. (4.19) as:

$$dD = d|\varepsilon^{pl}| \times \frac{\sigma^2}{2Er(1-D)^2}. \tag{4.21}$$

The last equation can be rearranged to give

$$\frac{dD}{d|\varepsilon^{pl}|} = \frac{\sigma^2}{2Er(1-D)^2}, \tag{4.22}$$

or

$$r = \frac{\sigma^2}{2E(1-D)^2 \frac{dD}{d|\varepsilon^{pl}|}}, \tag{4.23}$$

which allows to determine the material parameter r from a tensile test with reversed loading in the following way:

- Perform a uniaxial tensile test with reversed loading as shown in Fig. 4.10a. Determine the elastic modulus E of the undamaged material (initial modulus) and for each unloading-loading cycle n the modulus of the damaged material $\bar{E}_n$, the corresponding stress σ where the unloading starts and the plastic strain at $\sigma = 0$.
- Construct from the values of stress and corresponding plastic strain the flow curve as shown in Fig. 4.10b.
- Calculate from the modulus of the damaged material the damage variable D as given in Eq. (4.16). Plot the damage variable over the plastic strain as shown in Fig. 4.10c.
- Calculate the damage material parameter r according to Eq. (4.23) for given values of stress σ, elastic modulus E, damage variable D and slope $\frac{dD}{d|\varepsilon^{pl}|}$.
- Calculate an average (or interpolated) value of r by considering several evaluations as described in the previous step.

The extension to the three-dimensional case is in the following only very briefly sketched and the reader may refer to [40] for further details. The concept of effective stress and the hypothesis of strain equivalence remains valid and it can be stated based on the column matrices for stress and strain as:

$$\bar{\sigma} = \frac{1}{1-D}\sigma \text{ and } \bar{\varepsilon} = \varepsilon. \tag{4.24}$$

---

[4] August WÖHLER (1819–1914), German engineer.

(a)

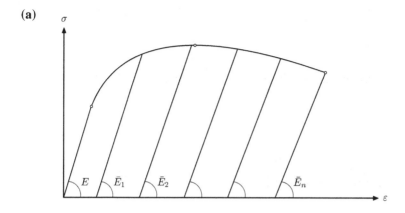

(b)

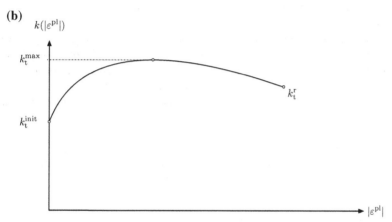

(c)

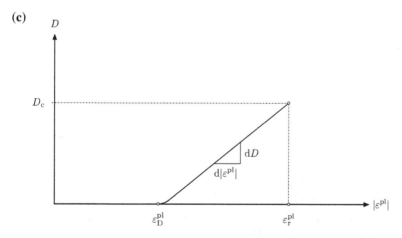

Fig. 4.10 **a** Uniaxial stress-strain diagram with reversed loading; **b** flow curve; **c** damage variable as a function of plastic strain

HOOKE's law in its elastic stiffness form (see Eq. (2.13)) reads now

$$\sigma = (1 - D)\varepsilon, \tag{4.25}$$

and the yield condition can be generally expressed as:

$$F(\sigma) = \frac{f(\sigma)}{1 - D} - k_t = 0. \tag{4.26}$$

4.3 Gurson's Damage Model

To model the arbitrary distribution of voids in a matrix (cf. Fig. 4.11a), GURSON introduced the idealized models of two void geometries [26]. For the first model, a single long circular cylindrical void is considered in a similarly shaped matrix (cf. Fig. 4.11b) whereas the second model considers a spherical shape of void and matrix (cf. Fig. 4.11c). The matrix material is considered as a homogeneous, isotropic, rigid-plastic VON MISES material. It should be highlighted here that the approach neglected the elastic material response and assumed the matrix material to be imcompressible.

Based on these assumptions, the following yield conditions could be derived

$$F = \left(\frac{\sigma_{eff}}{k_t}\right)^2 + 2D \cosh\left(\frac{3\sqrt{3}}{2}\frac{\sigma_m}{k_t}\right) - (1 + D^2) = 0, \text{ (cylindrical)} \tag{4.27}$$

$$F = \left(\frac{\sigma_{eff}}{k_t}\right)^2 + 2D \cosh\left(\frac{3}{2}\frac{\sigma_m}{k_t}\right) - (1 + D^2) = 0, \quad \text{(spherical)} \tag{4.28}$$

(a)   **(b)**   **(c)**

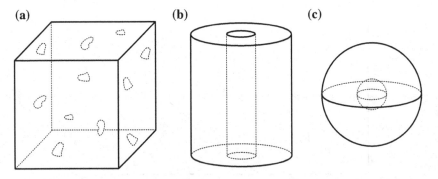

Fig. 4.11 **a** Voids of random shape and orientation distributed in a matrix; **b** long circular cylindrical void; **c** spherical void. Adapted from [26]

where σ_{eff} is the effective stress based on the VON MISES definition and σ_m is the mean stress (hydrostatic stress). The yield stress k_t is used to normalize the effective and mean stress. If the calculation of damage is based on the void volume fraction of real distributions as schematically shown in Fig. 4.11, then some account is taken of the interaction of neighboring voids [26].

Let us consider in the following the two-component σ-τ stress space. Thus, the yield conditions given in Eqs. (4.27) and (4.28) take the following form:

$$F = \left(\frac{\sqrt{\sigma^2 + 3\tau^2}}{k_t}\right)^2 + 2D\cosh\left(\frac{\sqrt{3}\,\sigma}{2\,k_t}\right) - (1 + D^2) = 0, \text{(cylindrical)} \quad (4.29)$$

$$F = \left(\frac{\sqrt{\sigma^2 + 3\tau^2}}{k_t}\right)^2 + 2D\cosh\left(\frac{1}{2}\frac{\sigma}{k_t}\right) - (1 + D^2) = 0. \quad \text{(spherical)} \quad (4.30)$$

A graphical representation of the GURSON yield condition based on Eq. (4.30) is shown in Fig. 4.12 where different values of the damage parameter have been assigned. A value of $D = 0$ results in the classical VON MISES ellipse and values

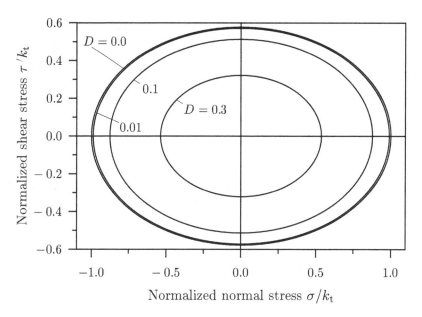

Fig. 4.12 Graphical representation of the yield condition according to GURSON in the two-component σ-τ space for different values of the damage variable (spherical voids assumed)

Table 4.3 Intersections of
the GURSON flow curves with
the coordinate axes

D	$\left(\frac{\tau}{k_t}\right)_{x=0}$	$\left(\frac{\sigma}{k_t}\right)_{y=0}$
0.00	±0.5774	±1.000000
0.01	±0.5715	±0.988639
0.10	±0.5132	±0.877740
0.30	±0.3215	±0.536890

$D > 0$ result in this graphical representation in similar looking shapes[5] but with a
smaller elastic region ($F < 0$), i.e. plastic yielding will start earlier compared to the
classical VON MISES condition.

Abscissae and ordinates of the intersection of the flow curve with the coordinate
axes can be obtained from Eq. (4.30) for $\sigma = 0$ or $\tau = 0$ and are summarized in
Table 4.3.

The derivative of the yield condition or the plastic strain increments with respect
to the stresses are obtained in the case of a general stress state from Eq. (4.28)

$$d\varepsilon^{pl} = d\lambda \frac{\partial F}{\partial \boldsymbol{\sigma}} = d\lambda \left(\frac{3 \mathbf{L} s}{k_t^2} + \frac{D}{k_t} \sinh\left(\frac{3\sigma_m}{2k_t}\right) \mathbf{1} \right), \tag{4.31}$$

where $s = \begin{bmatrix} s_x \; s_y \; s_z \; \tau_{xy} \; \tau_{yz} \; \tau_{xz} \end{bmatrix}^T$ is the column matrix of the stress deviator components[6], $\mathbf{L} = \lceil 1\,1\,1\,0\,0\,0 \rfloor$ is a diagonal scaling matrix and $\mathbf{1} = [1\,1\,1\,0\,0\,0]^T$ is
an identity column matrix. In the case of a σ-τ stress space, Eq. (4.30) gives under
consideration of $\frac{d}{dx}|x|^2 = 2|\sigma|\mathrm{sgn}(\sigma) = 2\sigma$ the following expression for the plastic
strain increments:

$$d\varepsilon^{pl} = \begin{bmatrix} d\varepsilon^{pl} \\ d\gamma^{pl} \end{bmatrix} = d\lambda \begin{bmatrix} \frac{2\sigma}{k_t^2} + \frac{D}{k_t} \sinh\left(\frac{\sigma}{2k_t}\right) \\ \frac{6\tau}{k_t^2} \end{bmatrix}. \tag{4.32}$$

At the end of this section, the evolution equation for the ductile damage should
be given. According to [40], the evolution equation is given as

$$dD = (1 - D)(d\varepsilon_x^{pl} + d\varepsilon_y^{pl} + d\varepsilon_z^{pl}) = (1 - D)d\varepsilon_V^{pl}, \tag{4.33}$$

where ε_V^{pl} is the volumetric plastic strain. Consideration of Eq. (4.31) and the fact
that $s_x + s_y + s_z = 0$ allows to reformulate the last equation to obtain:

$$dD = d\lambda \frac{3(D - D^2)}{k_t} \sinh\left(\frac{3\sigma_m}{2k_t}\right). \tag{4.34}$$

---

[5]The yield surfaces for $D > 0$ look like ellipses but because of the cosh function, they are not from
a mathematical point of view classified as ellipses.

[6]The first component of the stress deviator is given by $s_x = \frac{2}{3}\sigma_x - \frac{1}{3}(\sigma_y + \sigma_z)$.

Table 4.4 Basic equations of the original GURSON model (spherical inclusions, rigid-plastic material) in the case of a uniaxial stress state with σ as acting stress

1D GURSON model
Yield condition
$F = \left(\dfrac{\lvert\sigma\rvert}{k_{\mathrm{t}}}\right)^{2} + 2D\cosh\left(\dfrac{\sigma}{2k_{\mathrm{t}}}\right) - (1 + D^{2}) = 0$
Flow rule $\mathrm{d}\varepsilon^{\mathrm{pl}} = \mathrm{d}\lambda\left(\dfrac{2\sigma}{k_{\mathrm{t}}^{2}} + \dfrac{D}{k_{\mathrm{t}}}\sinh\left(\dfrac{\sigma}{2k_{\mathrm{t}}}\right)\right)$
Damage evolution equation
$\mathrm{d}D = \mathrm{d}\lambda\dfrac{3(D - D^{2})}{k_{\mathrm{t}}}\sinh\left(\dfrac{\sigma}{2k_{\mathrm{t}}}\right)$

Considering a pure one-dimensional stress state where only the normal stress σ is acting, Eqs. (4.30), (4.32) and (4.34) can be simplified to the forms given in Table 4.4.

In many practical applications, the GURSON yield condition is applied to *elasto-plastic* material behavior, even under consideration of isotropic hardening ($k_{\mathrm{t}} = k_{t}(\kappa)$). In such a case it is necessary to indicate the evolution equation for the internal variable κ. Assuming that the effective plastic strain is assigned as the internal variable, i.e. $\varepsilon_{\mathrm{eff}}^{\mathrm{pl}} = \kappa$, and furthermore assuming that the increment of equivalent plastic work in the matrix material equals the macroscopic increment of plastic work [68, 7], i.e.

$$\sigma\mathrm{d}\varepsilon^{\mathrm{pl}} = (1 - D)k_{\mathrm{t}}\mathrm{d}\varepsilon_{\mathrm{eff}}^{\mathrm{pl}}, \qquad (4.35)$$

the evolution of the internal variable is given in the one-dimensional case as:

$$\mathrm{d}\kappa = \frac{\mathrm{d}\lambda}{1 - D}\left(\frac{2\sigma^{2}}{k_{\mathrm{t}}^{3}} + \frac{D\sigma}{k_{\mathrm{t}}^{2}}\sinh\left(\frac{\sigma}{2k_{\mathrm{t}}}\right)\right). \qquad (4.36)$$

Table 4.5 Basic equations of the extended GURSON model (elastic range, isotropic hardening) in the case of a uniaxial stress state with σ as acting stress

1D GURSON model
HOOKE's law $\sigma = E\varepsilon$
Yield condition
$F = \left(\dfrac{\lvert\sigma\rvert}{k_{t}(\kappa)}\right)^{2} + 2D\cosh\left(\dfrac{\sigma}{2k_{t}(\kappa)}\right) - (1 + D^{2}) = 0$
Flow rule
$\mathrm{d}\varepsilon^{\mathrm{pl}} = \mathrm{d}\lambda\left(\dfrac{2\sigma}{(k_{t}(\kappa))^{2}} + \dfrac{D}{k_{t}(\kappa)}\sinh\left(\dfrac{\sigma}{2k_{t}(\kappa)}\right)\right)$
Evolution of hardening variable
$\mathrm{d}\kappa = \dfrac{\mathrm{d}\lambda}{1 - D}\left(\dfrac{2\sigma^{2}}{(k_{t}(\kappa))^{3}} + \dfrac{D\sigma}{(k_{t}(\kappa))^{2}}\sinh\left(\dfrac{\sigma}{2k_{t}(\kappa)}\right)\right)$
Evolution of damage variable
$\mathrm{d}D = \mathrm{d}\lambda\dfrac{3(D - D^{2})}{k_{t}(\kappa)}\sinh\left(\dfrac{\sigma}{2k_{t}(\kappa)}\right)$

Thus, the equations presented in Table 4.4 can be extended to the case of an *elasto-plastic* material with isotropic hardening as summarized in Table 4.5.

4.4 Supplementary Problems

4.2 Knowledge questions

- State the definition of the surface damage variable D.
- State the definition of the void volume fraction f.
- Explain the difference between the surface damage variable and the void volume fraction for real microstructures.
- Name different methods for the experimental determination of the damage parameter.
- How is HOOKE's law and the yield condition modified in the case of LEMAITRE's approach?
- State the assumptions in regards to the base material that GURSON made during the derivation of his yield conditions.
- State the different pore shapes that GURSON assumed during the derivation of his yield conditions.
- How is HOOKE's law of the macroscopic material modified in the set of equations used by GURSON?
- To which yield condition is the GURSON yield condition reduced for the special case of $D = 0.0$?

4.3 Uniaxial tensile test with and without consideration of contraction
State HOOKE's law for a uniaxial tensile test with and without consideration of contraction as shown in Fig. 4.9.

4.4 Uniaxial tensile test and volume change
Given is an idealized tensile sample with a rectangular cross section of $a \times a$. The length of the unloaded specimen is equal to $3a$. The specimen is elongated by a displacement of $\frac{a}{10}$ at both ends. Assume linear-elastic material behavior and calculate the volume change from the initial to the elongated state for a POISSON's ratio of 0.3.

4.5 Spherical RVE and pore—definition of damage
Given is a spherical RVE in the following. The pore itself is also modelled as a small sphere. The initial radius of the undamaged RVE is given as R_i. The damaged RVE contains a small spherical pore of radius r_D and the unknown outer radius is denoted by R. Determine based on the assumption of elastic material behavior and that no residual micro-stresses are acting on the damaged RVE the outer radius R. Then, calculate the damage parameter D and the void volume fraction f and related these results to the relative density $(\varrho - \varrho_i)/\varrho_i$.

4.6 Cubic RVE and spherical pore—definition of damage

Given is a cubic RVE in the follwoing. The pore itself is modelled as a small sphere. The initial side length of the undamaged RVE is given as a_i. The damaged RVE contains a small spherical pore of radius r_D and the unknown outer side length is denoted by a. Determine based on the assumption of elastic material behavior and that no residual micro-stresses are acting on the damaged RVE the outer side length a. Then, calculate the damage parameter D and the void volume fraction f and related these results to the relative density $(\varrho - \varrho_i)/\varrho_i$.

4.7 Cubic RVE and spherical pore—comparison between area and volume damage fraction

Consider a cubic RVE of side length $a = 1$ which contains in its center a spherical pore of diameter $D = 0.7$. Calculate and compare the volume (f) and area (D) damage variable. Evaluate the damaged area in a plane parallel to one of the side faces of the cube through the cube's center.

4.8 Damage evaluation from variation of elastic modulus

Given is an experimental stress-strain diagram for copper as shown in Fig. 4.13. At certain stress levels, the load was reversed to zero and then the specimen reloaded again. The exact stress-strain values of the reversal points are summarized in Table 4.6.

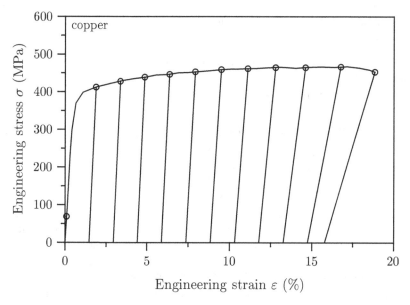

Fig. 4.13 Uniaxial stress-strain diagram with reversed loading for copper at room temperature. Data partially extracted from [32]

Table 4.6 Stress and strain values at the reversal points of the diagram shown in Fig. 4.13

Stress (MPa)	Total strain (%)	Plastic strain (%)
0	0	0
70.000	0.070707	0
412.105	1.900	1.482
428.300	3.377	2.941
438.862	4.868	4.420
447.261	6.374	5.894
452.681	7.923	7.367
461.379	9.502	8.841
460.805	11.108	10.314
462.781	12.785	11.788
464.804	14.600	13.261
465.761	16.757	14.735
452.906	18.816	15.766

Calculate and sketch the flow curve $k = k(|\varepsilon^{\text{pl}}|)$, the damage variable as a function of plastic strain $D = D(|\varepsilon^{\text{pl}}|)$ and the material parameter r in the evolution equation of the ductile damage, i.e. $r = \sigma^2/(2E(1 - D)^2 dD/d|\varepsilon^{\text{pl}}|)$.

4.9 Volumetric damage measurement based on Archimedes' principle

Derive an equation to experimentally determine the densities of small undamaged and damaged samples in order to calculate the volumetric damage variable. Base your approach on ARCHIMEDES' principle and consider weighting in air and immersing in a liquid.

4.10 Gurson's law: Difference between the cylindrical and spherical pore assumption

Consider the two-component σ-τ stress space for GURSON's law:

$$F = \left(\frac{\sqrt{\sigma^2 + 3\tau^2}}{k_t}\right)^2 + 2D\cosh\left(\frac{\sqrt{3}}{2}\frac{\sigma}{k_t}\right) - (1 + D^2) = 0, \text{(cylindrical)} \quad (4.37)$$

$$F = \left(\frac{\sqrt{\sigma^2 + 3\tau^2}}{k_t}\right)^2 + 2D\cosh\left(\frac{1}{2k_t}\frac{\sigma}{}\right) - (1 + D^2) = 0. \quad \text{(spherical)} \quad (4.38)$$

Calculate the maximum difference between both approaches for (a) a pure uniaxial stress state and (b) a pure shear stress state in the case of $D = 0.0; 0.01; 0.1; 0.3$.

4.11 Gurson's law: Expression for volumetric plastic strain

Show that the volumetric plastic strain $d\varepsilon_V^{\text{pl}}$ can be expressed in the case of GURSON's model as:

$$d\varepsilon_V^{\text{pl}} = d\lambda \frac{3D}{k_t} \sinh\left(\frac{3\sigma_m}{3k_t}\right). \quad (4.39)$$

Chapter 5
Fracture Mechanics

Abstract This chapter introduces the general concept of failure criteria in the presence of cracks. The topic is first introduced based on the concept of stress concentration and then generalized based on the stress intensity factor, the energy release rate, and the J-integral. The chapter introduces as well experimental approaches based on standardized specimen types to formulate the failure criteria.

5.1 Linear Fracture Mechanics

5.1.1 Stress Concentration

Let us consider an infinite plate with a circular hole of diameter $2a$ which is loaded in one direction by a normal stress σ, see Fig. 5.1. The location of a small element of this plate can be described based on the polar coordinates (r, φ) with respect to the hole's center. Let us assume in the following a plane stress state.

The three stress components, i.e. the normal stress component in the radial direction σ_r, the normal stress component in the circumferential direction σ_φ and the shear stress component $\tau_{r\varphi}$, are given by [66]

$$\sigma_r(r, \varphi) = \frac{\sigma}{2}\left(1 - \frac{a^2}{r^2}\right) + \frac{\sigma}{2}\left(1 + \frac{3a^4}{r^4} - \frac{4a^2}{r^2}\right)\cos(2\varphi)\,, \tag{5.1}$$

$$\sigma_\varphi(r, \varphi) = \frac{\sigma}{2}\left(1 + \frac{a^2}{r^2}\right) - \frac{\sigma}{2}\left(1 + \frac{3a^4}{r^4}\right)\cos(2\varphi)\,, \tag{5.2}$$

$$\tau_{r\varphi}(r, \varphi) = -\frac{\sigma}{2}\left(1 - \frac{3a^4}{r^4} + \frac{2a^2}{r^2}\right)\sin(2\varphi)\,. \tag{5.3}$$

© Springer Science+Business Media Singapore 2016
A. Öchsner, *Continuum Damage and Fracture Mechanics*,
DOI 10.1007/978-981-287-865-6_5

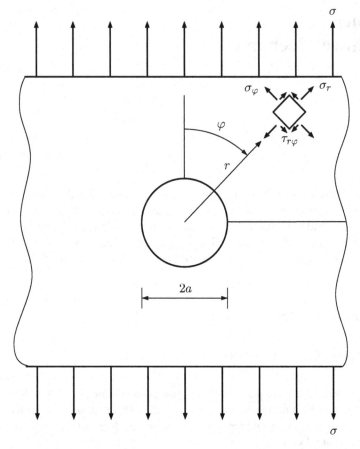

Fig. 5.1 Infinite plate with a circular hole

Along the minimum cross section, i.e. for $\varphi = 90° \vee 270°$, Eqs. (5.1) to (5.3) reduce to:

$$\sigma_r(r) = \frac{\sigma}{2}\left(\frac{3a^2}{r^2} - \frac{3a^4}{r^4}\right), \tag{5.4}$$

$$\sigma_\varphi(r) = \frac{\sigma}{2}\left(\frac{3a^4}{r^4} + \frac{a^2}{r^2} + 2\right), \tag{5.5}$$

$$\tau_{r\varphi}(r) = 0. \tag{5.6}$$

The graphical representation of $\sigma_r(r, \varphi = 0)$ and $\sigma_\varphi(r, \varphi = 0)$ is given in Fig. 5.2. It can be seen from this figure that the dominant stress component is σ_φ which takes at $r = a$ three times the value of the external load σ. At sufficient distance from the hole, the circumferential stress decays to the value of the external load, i.e. $\sigma_\varphi = \sigma$.

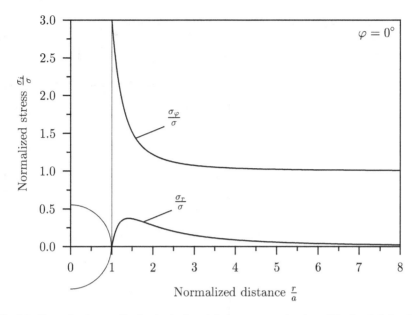

Fig. 5.2 Normalized stress distribution in the minimum cross section ($\varphi = 0°$) of an infinite plate with a circular hole

The increase of the circumferential stress component towards the root of the hole is called stress concentration. To characterize this stress increase, the so-called stress concentration factor K_t is introduced[1] as the ratio between the peak stress σ_{max} to the nominal stress σ_{nom}:

$$K_t = \frac{\sigma_{max}}{\sigma_{nom}}, \tag{5.7}$$

where the nominal stress is the stress that would exist in the plate if the hole would not be existing (in the case of a finite plate, the overall section would be used to calculate this value) or alternatively in the case of a finite plate the stress value which results from the nominal net cross section.

Let us investigate now the influence of the finite width of a plate on the stress concentration, see Fig. 5.3. Assuming a width of $2A$, a hole diameter of $2a$ and a thickness of t, the macroscopic stress σ due to an external force F can be calculated based on

$$\sigma = \frac{F}{2At}, \tag{5.8}$$

---

[1]Sometimes called the theoretical stress concentration factor [45].

Fig. 5.3 Semi-infinite plate
with a circular hole

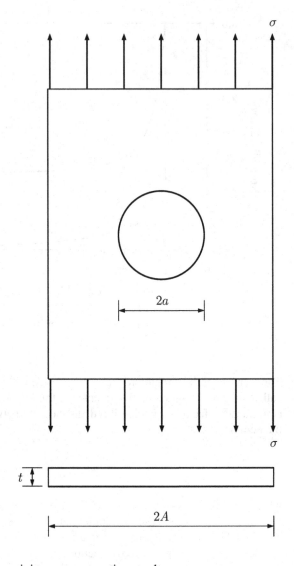

while the nominal stress in the minimum cross section results as:

$$\sigma_{\text{nom}} = \frac{F}{2(A - a)t}. \tag{5.9}$$

Based on the definition of the nominal stress in Eq. (5.9), the stress concentration for
a semi-infinite plate (see Fig. 5.3) is given as [45]:

$$K_{\text{t}} = \frac{\sigma_{\max}}{\sigma_{\text{nom}}} = 3.0 - 3.140 \left(\frac{a}{A} \right)^1 + 3.667 \left(\frac{a}{A} \right)^2 - 1.527 \left(\frac{a}{A} \right)^3 \quad \text{for } 0 \leq \frac{a}{A} \leq 1. \tag{5.10}$$

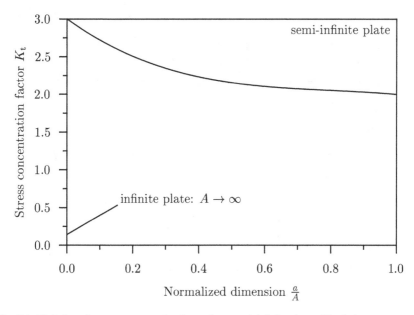

Fig. 5.4 Variation of stress concentration factor for a semi-infinite plate with a hole

It can be seen from Eq. (5.10) that the limiting case of an infinite plate, i.e. $K_t(A \rightarrow \infty) = 3$, is included. The distribution of the stress concentration factor in the given range of the major dimensions, i.e. $0 \leq \frac{a}{A} \leq 1$, is shown in Fig. 5.4. It can be seen that the influence of the finite width reduces the ideal stress concentration of 3 (infinite plate), converging to a minimum value of 2.

Let us look now at the case of an infinite plate with an elliptical hole where the minor axis is parallel to the loading direction, see Fig. 5.5.

The maximum stress in the root of the elliptical hole is given by [25, 45]

$$\sigma_{max} = \sigma\left(1 + 2\frac{a}{b}\right). \tag{5.11}$$

If $a = b$, the classical solution for the circular hole is obtained, see Eq. (5.5) for $r = a$. Let us now imagine that the ellipse becomes flatter and flatter or in other words that $b \ll a$. This means that the stress according to Eq. (5.11) becomes higher and higher and reaches in the limiting case, i.e. $b \rightarrow 0$, infinity. In this case, the transition from a hole to a crack is obtained, see Fig. 5.6.

5.1 Example: Semi-infinite plate with a hole—elasto-plastic material behavior
Consider a semi-infinite plate as schematically shown in Fig. 5.3. The external load is given by a single force of $F = 60$ kN. The geometrical dimensions are as follows: Hole diameter $2a = 20$ mm, plate width $2A = 100$ mm, and plate thickness $t = 10$ mm. Consider a safety factor of 2 and assume an aluminum alloy with an initial

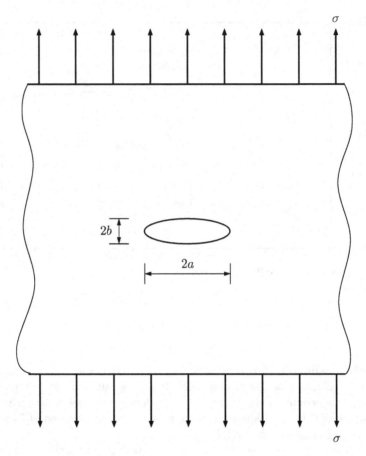

Fig. 5.5 Infinite plate with an elliptical hole

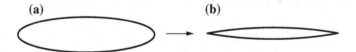

Fig. 5.6 Transition from **a** hole to **b** crack

yield tress of $k_t^{\text{init}} = 350$ MPa. Check based on the VON MISES yield condition if plastic deformation occurs in the region far from the hole or in the minimum cross section.

5.1 Solution
The macroscopic stress is obtained as:

$$\sigma = \frac{2F}{2At} = \frac{2 \times 60000}{100 \times 10} \text{ MPa} = 120 \text{ MPa}. \qquad (5.12)$$

This stress is representative for a region far from the hole. Since the stress state is uniaxial far from the hole, this value can be directly compared to the initial yield stress. Thus, this region is in the elastic range.

The nominal stress is based on the minimum cross section and can be calculated according to:

$$\sigma_{nom} = \frac{2F}{2(A-a)t} = \frac{2 \times 60000}{2 \times (50-10) \times 10} \, \text{MPa} = 150 \, \text{MPa}. \tag{5.13}$$

This value is again smaller than the initial yield stress. It should be noted here that this stress is assumed in the case of damage mechanics, see Chap. 4. Let us now consider the stress concentration according to Eq. (5.10) with $\frac{a}{A} = \frac{1}{5}$:

$$K_t = 3.0 - 3.140 \left(\frac{1}{5}\right)^1 + 3.667 \left(\frac{1}{5}\right)^2 - 1.527 \left(\frac{1}{5}\right)^3 = 2.506. \tag{5.14}$$

It can be now concluded that the maximum stress is obtained as:

$$\sigma_{max} = K_t \times \sigma_{nom} = 2.506 \times 150 \, \text{MPa} = 376 \, \text{MPa}. \tag{5.15}$$

This value clearly exceeds the initial yield stress. It should be noted here that the stress state in the root of the hole ($r = a; \varphi = 90°$) is purely uniaxial. However, there is a circumferential (σ_φ) and radial (σ_r) stress acting in the minimum cross section, see Fig. 5.2. The question is now if an interaction of the circumferential and the radial stress components according to the VON MISES yield condition results in another maximum than ($r = a; \varphi = 90°$). The general formulation of the yield condition was given in Eq. (3.91)

$$F = \underbrace{\sqrt{\frac{1}{2}\left((\sigma_x - \sigma_y)^2 + (\sigma_y - \sigma_z)^2 + (\sigma_z - \sigma_x)^2\right) + 3\left(\sigma_{xy}^2 + \sigma_{yz}^2 + \sigma_{xz}^2\right)}}_{\sigma_{eff}} - k_t.$$

$$\tag{5.16}$$

Modifying this formulation according to our situation, i.e. $\sigma_x \rightarrow \sigma_\varphi, \sigma_y \rightarrow \sigma_r$, gives:

$$F = \sqrt{\sigma_\varphi^2 + \sigma_r^2 - \sigma_\varphi \sigma_r} - k_t^{init} = 0. \tag{5.17}$$

The stress components σ_φ and σ_r together with the effective VON MISES stress according to Eq. (5.17) are shown in Fig. 5.7. It can be seen from this figure that the stress maximum remains at $r = a$ and that the interaction of the stress components reduces the effective stress σ_{eff} compared to the circumferential component σ_φ.

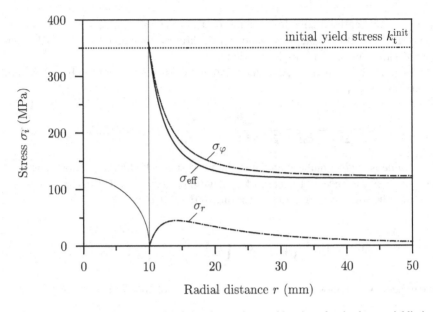

Fig. 5.7 Stress distribution in a semi-infinite plate under consideration of a plastic material limit

5.2 Example: Semi-infinite plate with a hole—elasto-plastic material behavior under consideration of different material behavior

Reconsider Ex. 5.1 and assume now that the plate is made of the titanium alloy Ti-6Al-4V. The initial tensile yield stress for this material is $k_t^{init} = 910$ MPa. Check if any location of the plate undergoes plastic deformation and draw conclusions on the maximum stress and yield behavior in dependance of the introduced material. The geometry and the boundary condition remain unchanged.

5.2 Solution

The previous example clarified that the critical location is in the root of the hole. Macroscopic stress, nominal stress and the stress concentration factor do not change since these quantities do not depend on the material properties. This statement is correct under the assumption that the external load for both cases is the same and given as a force. Thus, the maximum stress is 376 MPa which acts at the root of the hole. Since only a single stress component is acting in the critical location, the stress of 376 MPa can be directly compared to the intial yield stress (910 MPa) and it can be concluded that the entire plate is in the elastic range.

5.1.2 Stress Intensity—Crack-Tip Field

The theory of fracture mechanics distinguishes three basic crack opening modes, see Fig. 5.8.

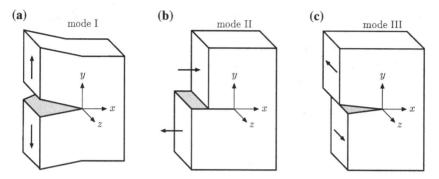

Fig. 5.8 Crack opening modes: **a** opening mode, **b** sliding mode and **c** tearing mode. Adapted from [60]

Fig. 5.9 Coordinate systems at the crack front

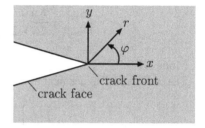

Mode I or the opening mode (see Fig. 5.8a) is characterized by opening with respect to the x-z plane. Mode II or the sliding mode (see Fig. 5.8b) is characterized by a sliding in x-direction. Mode III or the tearing mode (see Fig. 5.8c) is characterized by relative displacements in the z-direction. However, mode I is considered the most important case from a practical point of view. The following part summarizes the stress and displacement field for the different modes under the assumption of a two-dimensional problem and a straight crack [24]. The relationships are valid for the crack-tip field, i.e. the area close to the crack tip. The in Fig. 5.9 introduced polar coordinate system (r, φ) is used for convenience.

- **Mode I**
  The stress and displacement field can be written as:

$$
\begin{bmatrix} \sigma_x \\ \sigma_y \\ \tau_{xy} \end{bmatrix} = \frac{K_{\mathrm{I}}}{\sqrt{2\pi r}} \cos \tfrac{\varphi}{2} \begin{bmatrix} 1 - \sin \tfrac{\varphi}{2} \sin \tfrac{3\varphi}{2} \\ 1 + \sin \tfrac{\varphi}{2} \sin \tfrac{3\varphi}{2} \\ \sin \tfrac{\varphi}{2} \cos \tfrac{3\varphi}{2} \end{bmatrix}, \tag{5.18}
$$

$$
\begin{bmatrix} u_x \\ u_v \end{bmatrix} = \frac{K_{\mathrm{I}}}{2G} \sqrt{\frac{r}{2\pi}} (\kappa - \cos \varphi) \begin{bmatrix} \cos \tfrac{\varphi}{2} \\ \sin \tfrac{\varphi}{2} \end{bmatrix}, \tag{5.19}
$$

where the stress intensity factor K_I characterizes the 'strength' (amplitude) of the fields. This stress intensity factor depends on the geometry of the specimen and crack as well as on the type of loading. The factor κ allows to distinguish between the plane strain ($\kappa = 3 - 4\nu$) and the plane stress ($\kappa = (3 - \nu)/(1 + \nu)$) case.

- **Mode II**
  The stress and displacement field can be written as:

$$\begin{bmatrix} \sigma_x \\ \sigma_y \\ \tau_{xy} \end{bmatrix} = \frac{K_{II}}{\sqrt{2\pi r}} \begin{bmatrix} -\sin \frac{\varphi}{2} \left[2 + \cos \frac{\varphi}{2} \cos \frac{3\varphi}{2} \right] \\ \sin \frac{\varphi}{2} \cos \frac{\varphi}{2} \cos \frac{3\varphi}{2} \\ \cos \frac{\varphi}{2} \left[1 - \sin \frac{\varphi}{2} \sin \frac{3\varphi}{2} \right] \end{bmatrix}, \qquad (5.20)$$

$$\begin{bmatrix} u_x \\ u_v \end{bmatrix} = \frac{K_{II}}{2G} \sqrt{\frac{r}{2\pi}} \begin{bmatrix} \sin \frac{\varphi}{2} [\kappa + 2 + \cos \varphi] \\ \cos \frac{\varphi}{2} [\kappa - 2 + \cos \varphi] \end{bmatrix}. \qquad (5.21)$$

- **Mode III**
  The stress and displacement field can be written as:

$$\begin{bmatrix} \tau_{xz} \\ \tau_{yz} \end{bmatrix} = \frac{K_{III}}{\sqrt{2\pi r}} \begin{bmatrix} -\sin \frac{\varphi}{2} \\ \cos \frac{\varphi}{2} \end{bmatrix}, \qquad (5.22)$$

$$u_z = \frac{2K_{III}}{2G} \sqrt{\frac{r}{2\pi}} \sin \frac{\varphi}{2}. \qquad (5.23)$$

It can be seen from the above equations that the stresses at the crack tip have a singularity of the type $r^{-1/2}$ whereas the displacement fields show a behavior of the type $r^{1/2}$. Let us summarize at this point the principal difference between classical mechanics and the consideration of a stress concentration or singularity, see Fig. 5.10.

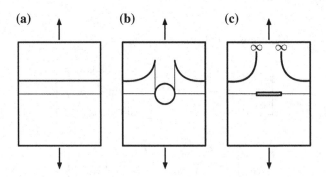

Fig. 5.10 Stress distribution in the case of **a** classical mechanics, **b** concentration and **c** singularity

Classical mechanics predicts in the case of a flat plate approximately a constant stress distribution in the cross section (see Fig. 5.10a). Introducing a hole (see Fig. 5.10b) results in an increasing stress towards the root of the hole and in the case of a circular hole, a stress concentration factor of approx. 3 is observed. Degenerating the hole to a crack gives a singularity where the stress converges against infinity (see Fig. 5.10b). As a result of this infinite stress, the concepts of classical plasticity theory (see Chap. 3) cannot be applied since a comparison of an infinite stress with a stress limit is meaningless. Different concepts, e.g. based on the stress intensity factor, must be applied to judge the failure of a material.

Table 5.1 summarizes for some simple geometries and load cases the stress intensity factor K_{I}. It can be seen from this table that the expressions for the stress intensity factor can be generalized as

$$K_{\mathrm{I}} = \sigma \sqrt{\pi a} \times Y_{\mathrm{c}}, \tag{5.24}$$

where the dimensionless parameter Y_{c} depends on the specimen and crack geometry as well as the loading conditions.

Based on this stress intensity factor, the following failure criterion can be formulated for the mode I crack opening mode.

$$K_{\mathrm{I}} = K_{\mathrm{Ic}}, \tag{5.25}$$

where the so-called fracture toughness K_{Ic} is a critical material parameter. Ranges of the fracture toughness for some common classes of engineering materials are collected in Table 5.2. It should be noted here that the fracture toughness has the uncommon unit $\mathrm{MPa}\sqrt{\mathrm{m}}$. Further solutions for the stress intensity factor can be found in the reference books [4, 37, 58, 62].

In the case of pure mode II or III crack opening modes, one can formulate, similar to Eq. (5.26), the following failure criteria:

$$K_{\mathrm{II}} = K_{\mathrm{IIc}}, \tag{5.26}$$
$$K_{\mathrm{III}} = K_{\mathrm{IIIc}}. \tag{5.27}$$

In the case that the different crack opening modes are acting simultaneously, a generalized failure criterion of the form

$$f(K_{\mathrm{I}}, K_{\mathrm{II}}, K_{\mathrm{III}}) = 0, \tag{5.28}$$

can be formulated and one speaks of mixed-mode loading, see [25]. In the case of the same structural and crack geometry, different load cases for the same mode of opening can be considered by adding stress-intensity factor solutions. This is known as the principle of superposition.

Table 5.1 Selected stress intensity factors for simple geometries and mode I load cases

Configuration	Stress intensity factor
	$K_{\mathrm{I}} = \sigma\sqrt{\pi a}$
	$K_{\mathrm{I}} = \sigma\sqrt{\pi a}\, F_{\mathrm{I}}$, $F_{\mathrm{I}} = \dfrac{1-0.025(a/b)^2+0.06(a/b)^4}{\sqrt{\cos(\pi a/2b)}}$
	$K_{\mathrm{I}} = 1.1215\,\sigma\sqrt{\pi a}$
	$K_{\mathrm{I}} = \sigma\sqrt{\pi a}\sqrt{\dfrac{2b}{\pi a}\,\tan\dfrac{\pi a}{2b}}\,G_{\mathrm{I}}$, $G_{\mathrm{I}} = \dfrac{0.752+2.02\frac{a}{b}+0.37\left(1-\sin\frac{\pi a}{2b}\right)^3}{\cos\frac{\pi a}{2b}}$
	$K_{\mathrm{I}} = \sigma\sqrt{\pi a}\sqrt{\dfrac{2b}{\pi a}\,\tan\dfrac{\pi a}{2b}}\,G_{\mathrm{I}}$, $G_{\mathrm{I}} = \dfrac{0.923+0.199\left(1-\sin\frac{\pi a}{2b}\right)^4}{\cos\frac{\pi a}{2b}}$

Adapted from [25]

Table 5.2 Initial yield stress and fracture toughness of some engineering materials

Material	k_t^{init} (MPa)	K_{Ic} (MPa$\sqrt{\text{m}}$)
High-strength steel	1600...2000	25...95
Construction steel	<500	30...125
Ti alloys	800...1200	40...95
Al alloys	200...600	20...65
PMMA	53.8...73.1	0.7...1.6

Adapted from [25, 13]

5.3 Example: Infinite plate with a crack—critical stress

Consider a small crack in a thin infinite plate as schematically shown in Fig. 5.11. The plate is made of the steel alloy 4340 with the following material properties: $k_t^{\text{init}} = 1420$ MPa and $K_{\text{Ic}} = 87.4$ MPa$\sqrt{\text{m}}$. The small crack has a total length of $2a = 0.5$ mm. Investigate if the specimen will fail under an external boundary stress of 1000 MPa.

5.3 Solution

The boundary stress is smaller than the initial yield stress. Thus, there is no global plastic deformation which could result in the failure of the plate. In the case under consideration, the stress intensity factor can be taken from Table 5.1 as:

$$K_{\text{I}} = \sigma\sqrt{\pi a} \,. \tag{5.29}$$

If the stress intensity factor K_{I} equals the fracture toughness K_{Ic}, then the critical stress is obtained, i.e. $\sigma \to \sigma_c$:

Fig. 5.11 Infinite plate with a small crack

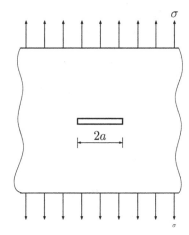

$$\sigma_c = \frac{K_{Ic}}{\sqrt{\pi a}} = \frac{87.4 \text{ MPa}\sqrt{m}}{\sqrt{\pi \, 0.5 \times 10^{-3}/2 \text{ m}}} = 3119 \text{ MPa}. \qquad (5.30)$$

Since $\sigma < \sigma_c$ holds, it can be concluded that no failure will occur.

5.4 Example: Engineering structure with a crack—failure estimate

Consider an engineering structure which is approximated as a thin plate with a small crack of size $2a$. The utilized material is the aluminium alloy 7075 with the following material properties: $k_t^{init} = 495$ MPa and $K_{Ic} = 24.0$ MPa$\sqrt{m}$. An experimental investigation revealed that the plate fails under an external boundary stress of 450 MPa for which the critical crack length is $2a = 2.5$ mm. Determine if the same engineering structure (assume mode I for both cases) will fail if an external boundary stress of $\sigma = 225$ MPa is applied and the crack has an actual length of $2a = 5$ mm, i.e. the load is divided in half while the crack doubled its length.

5.4 Solution

The dimensionless parameter Y_c can be determined from the first set of values:

$$Y_c = \frac{K_{Ic}}{\sigma\sqrt{\pi a}} = \frac{24.0 \text{ MPa}\sqrt{m}}{450 \text{ MPa}\sqrt{\pi \, 2.5 \times 10^{-3}/2 \text{ m}}} = 0.851077. \qquad (5.31)$$

The second set of values allows to calculate the critical stress as:

$$\sigma_c = \frac{K_{Ic}}{Y_c\sqrt{\pi a}} = \frac{24.0 \text{ MPa}\sqrt{m}}{0.851077\sqrt{\pi \, 5.0 \times 10^{-3}/2 \text{ m}}} = 318.2 \text{ MPa}. \qquad (5.32)$$

Since $\sigma = 225$ MPa $< \sigma_c = 318.2$ MPa holds, it can be concluded that no failure will occur.

5.5 Example: Edge crack structure loaded with tensile and bending loads (mode I)

An edge crack of length a is subjected to a combination of tensile and bending loads as shown in Fig. 5.12. Calculate the total stress intensity factor.

5.5 Solution

The principle of superposition is schematically shown in Fig. 5.13. The total stress intensity factor can be obtained from Table 5.1 as:

$$K_I = K_{I,t} + K_{I,b}. \qquad (5.33)$$

5.1.3 Test Methods for the Determination of the Fracture Toughness

The experimental determination of the fracture toughness K_{Ic} based on standard test specimens is described in the standard ASTM E399 [3] or, for example, in the

Fig. 5.12 Edge crack under combined loading

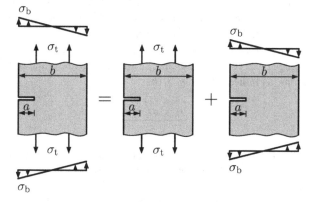

Fig. 5.13 Edge crack under combined loading—principle of superposition

textbook [28]. This method provides the *plane-strain* fracture toughness. Independent of the specific specimen geometry, a defined starter notch is machined and then a fatigue crack is produced by cyclic loading the notched specimen for a number of cycles, usually between about 10^4 and 10^6, see Fig. 5.14. The specimen crack size a is required for the test evaluation and indicated in Fig. 5.14b.

Fig. 5.14 a Starter notch (machined) and **b** fatigue crack (cyclic loading)

(a)   **(b)**

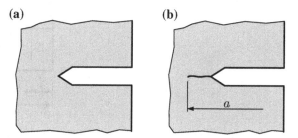

Fig. 5.15 Measured
variables (F, V) during test
execution

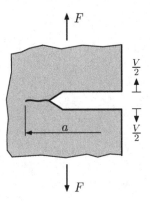

The subsequent test procedure is a quasi-static test at which the applied force F and the crack mouth opening displacement (CMOD) V (see Fig. 5.15) is recorded.

It should be highlighted here that the crack size a (see Figs. 5.14b and 5.15) is not measured during the quasi-static test. This dimension is evaluated after fracture based on the average of three measurements, i.e. at mid-thickness and at both quarter-thickness locations, as indicated in Fig. 5.16:

$$a = \frac{a_1 + a_2 + a_3}{3}.$$ (5.34)

The required force measurement is normally straightforward, i.e. the output of the force-sensing transducer (load cell) while the the crack opening (CMOD) requires, in general, more efforts.

The common procedure for the displacement measurement is the use of a so-called extensometer (clip gage), see Fig. 5.17. This extensometer measures the displacement of two precisely located gage positions spanning the crack starter notch mouth. The basic principle of measurement is, for example, based on the deformation of a small beam at which the strain is measured at the outer surface.

Fig. 5.16 Determination of
the crack length a. Adapted
from [9]

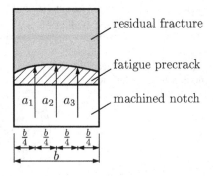

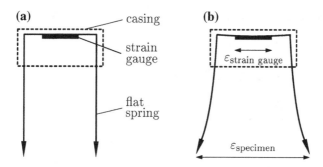

Fig. 5.17 Schematic sketch of an extensometer: **a** undeformed and **b** deformed

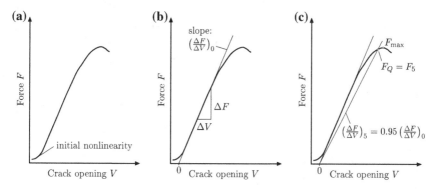

Fig. 5.18 Schematic force versus crack opening displacement curve: **a** recorded data, **b** slope in the linear part and **c** secant line through origin (0) with a slope of 95 % of original slope. Adapted from [9, 3]

A typical force-displacement recording for a fracture toughness test is shown in Fig. 5.18a. The test evaluation requires now to draw the tangent to the linear portion of the record, see Fig. 5.18b. The slope of this tangent is denoted by $\left(\frac{\Delta F}{\Delta V}\right)_0$. The next step is to draw the secant line through the origin 0 with a slope equal to $\left(\frac{\Delta F}{\Delta V}\right)_5 = 0.95 \left(\frac{\Delta F}{\Delta V}\right)_0$. This secant intersects the graph of the recorded data at $F = F_5$.

The conditional force F_Q, which is used to calculate a conditional fracture toughness K_Q (see the following Tables 5.3, 5.4, 5.5 and 5.6 and replace F by F_Q), is obtained depending on the graph shape as follows: if the force at every point on the record which precedes P_5 is lower than P_5 (see Fig. 5.19a, type 1), then P_5 is P_Q. If, however, there is a maximum force preceding P_5 which exceeds it (see Fig. 5.19b and c, type 2 and 3), then this maximum is P_Q.

To judge if the conditional fracture toughness K_Q is equal to the fracture toughness, i.e.

Table 5.3 Standard bend specimen SE (B)

Specimen type

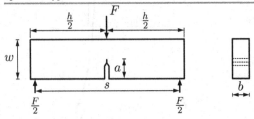

Conditional equation	Dimension
$K_{\mathrm{I}} = \dfrac{F s}{b w^{3/2}} f_1(a/w)$	$w = 2b$, standard specimen
$f_1(a/w)$ geometric factor	$w = b$ to $4b$, alternative specimen
(cf. Table 5.7)	$s = 4w$
for $0.45 \leq a/w \leq 0.55$	$h = \text{min. } 4.2w$
and $s/w = 4$	$a = (0.45 \ldots 0.55)w$

Adapted from [9, 3]

Table 5.4 Standard compact specimen C (T)

Specimen type

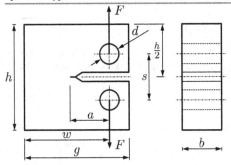

Conditional equation	Dimension
$K_{\mathrm{I}} = \dfrac{F}{b w^{1/2}} f_2(a/w)$	$w = 2b$, standard specimen
$f_2(a/w)$ geometric factor	$w = 2b$ to $4b$, alternative specimen
(cf. Table 5.7)	$s = 0.55w$
for $0.45 \leq a/w \leq 0.55$	$h = 1.2w$
	$a = (0.45 \ldots 0.55)w$
	$d = 0.25w$
	$g = 1.25w$

Adapted from [9, 3]

Table 5.5 Arc-shaped tension specimen A (T)

Specimen type

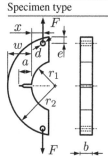

Conditional equation	Dimension
$K_{\mathrm{I}} = \dfrac{F}{bw^{1/2}}\, f_3(a/w)\left[\dfrac{3x}{w} + 1.9 + 1.1\left(\dfrac{a}{w}\right)\right]$	$w = 2b$, standard specimen
$\times \left[1 + 0.25(1 - a/w)^2\left(1 - \dfrac{r_1}{r_2}\right)\right]$	$w = 2b$ to $4b$, alternative specimen
$f_3(a/w)$ geometric factor	$d = 0.25w$
(cf. Table 5.7)	$e = 0.25w$
for $0.45 \le a/w \le 0.55$	$a = (0.45\ldots0.55)w$
	$x = 0.50w$

Adapted from [9, 3]

$$K_{\mathrm{Q}} \overset{?}{=} K_{\mathrm{Ic}}, \qquad (5.35)$$

two conditions must be verified:

- Plane-strain condition:

$$b, a \ge 2.5\left(\frac{K_{\mathrm{Q}}}{k_{\mathrm{t}}^{\mathrm{init}}}\right)^2, \qquad (5.36)$$

where the initial yield strength is defined as the 0.2 % offset value. If this relationship is fulfilled, then K_{Q} is equal to K_{Ic}. Otherwise, the test is not valid and a larger specimen size is required to determine the fracture toughness. The new specimen thickness b can be estimated from Eq. (5.36) and should be at least 50 % larger than the original thickness.

- Maximum stress intensity during fatigue loading:
  The maximum stress-intensity factor $K_{\max}$ during any stage of the fatigue crack growth should fulfill the following condition:

$$K_{\max} \le 0.6 K_{\mathrm{Q}}. \qquad (5.37)$$

Table 5.6 Standard disk-shaped compact specimen DC (T)

Specimen type

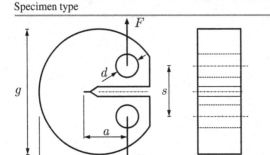

Conditional equation	Dimension
$K_{\mathrm{I}} = \dfrac{F}{bw^{1/2}}\, f_4(a/w)$	$w = 2b$, standard specimen
$f_4(a/w)$ geometric factor	$w = 2b$ to $4b$, alternative specimen
(cf. Table 5.7)	$s = 0.55w$
for $0.45 \leq a/w \leq 0.55$	$a = (0.45 \ldots 0.55)w$
	$g = 1.35w$
	$d = 0.25w$
	$x = 0.25w$

Adapted from [9, 3]

Let us now have a look on the different standard specimen geometries. The standard bend specimen (SE (B)) is a beam loaded in three-point bending, see Table 5.3. This specimen requires the lowest loads to conduct the test.

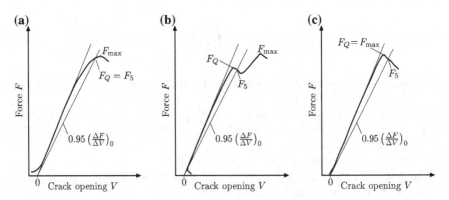

Fig. 5.19 Principal types of force-displacement records: **a** type 1, **b** type 2 and **c** type 3. Adapted from [9, 3]

Table 5.7 Geometric factor $f_i(a/w)$ $(i = 1 \ldots 4)$ for the conditional equations in Tables 5.3, 5.4, 5.5 and 5.6 for $0.45 \leq a/w \leq 0.55$

a/w	$f_1(a/w)$ SE (B)	$f_2(a/w)$ C (T)	$f_3(a/w)$ A (T)	$f_4(a/w)$ DC (T)
0.450	2.29	8.34	3.23	8.71
0.455	2.32	8.46	3.27	8.84
0.460	2.35	8.58	3.32	8.97
0.465	2.39	8.70	3.37	9.11
0.470	2.43	8.83	3.42	9.25
0.475	2.46	8.96	3.47	9.40
0.480	2.50	9.09	3.52	9.55
0.485	2.54	9.23	3.57	9.70
0.490	2.58	9.37	3.62	9.85
0.495	2.62	9.51	3.68	10.01
0.500	2.66	9.66	3.73	10.17
0.505	2.70	9.81	3.79	10.34
0.510	2.75	9.96	3.85	10.51
0.515	2.79	10.12	3.91	10.68
0.520	2.84	10.29	3.97	10.86
0.525	2.89	10.45	4.03	11.05
0.530	2.94	10.63	4.10	11.24
0.535	2.99	10.80	4.17	11.43
0.540	3.04	10.98	4.24	11.63
0.545	3.09	11.17	4.31	11.83
0.550	3.14	11.36	4.38	12.04

Adapted from [9, 3]

The compact tension (C (T)) specimen is characterized by a lower material volume (for the same specimen thickness b) compared to the bend specimen, see Table 5.4.

The arc-shaped tension specimen (A (T)) was designed to obtain values for thick high-pressure cylinders, see Table 5.5.

The disk-shaped compact specimen (DC (T)) can be economically machined from round semifinished parts, see Table 5.6.

Looking at the specimen geometries and the experimental procedure, which may require iterations, it can be stated that the determination of the fracture toughness is costly and time-consuming.

Let us highlight here that the fracture toughness has a strong dependency on the specimen thickness, i.e. if the stress state is plane stress or plane strain, and the temperature, see Fig. 5.20.

It is important to highlight a few issues at the end of this section, see Fig. 5.21. All the previous derivations were done under the assumption of pure elastic material behavior, i.e. linear elasticity theory. However, in-situ SEM investigations confirm the existence of a small process zone where microcracks and voids develop and

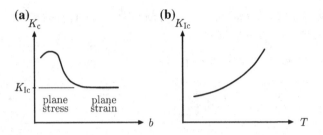

Fig. 5.20 Dependency of fracture toughness from **a** specimen thickness and **b** temperature. Adapted from [25]

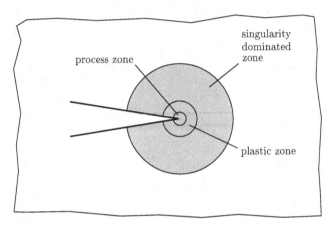

Fig. 5.21 Different zones around the crack tip. Adapted from [25]

finally coalesce with the main crack [12]. Furthermore, the assumption of pure elastic material behavior is no more fulfilled at a certain vicinity to the crack tip and a plastic zone is developed. The concept of stress intensity factors is valid in this singularity dominated zone. Within this theory, it is assumed that this zone is much larger than the two inner regions and that this zone even controls the small inner regions.

5.1.4 Energy Release Rate

Let us look in the following on two basic mechanical members, i.e. a spring and a bar as shown in Fig. 5.22. Under the influence of an external force F, both members elongate by a displacement of u.

The elastic strain energy or deformation energy W_{int} of both systems can be generally expressed as

$$W_{\text{int}} = \int_0^u F \, d\hat{u} . \tag{5.38}$$

Fig. 5.22 Simple
mechanical members:
a spring and **b** bar

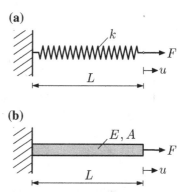

Assuming linear-elastic material behavior, i.e. $F = ku$ or $\sigma = E\varepsilon$, the elastic strain energy can be expressed as

$$W_{\text{int}} = \int_0^u k\hat{u}\, d\hat{u} = \frac{1}{2}ku^2 = \frac{1}{2}Fu\,, \tag{5.39}$$

where k is the spring constant. It should be noted here that a spring and a bar are from a mechanical point of view identical members if $k = \frac{EA}{L}$ holds for the bar. For some application, the volumetric strain energy is used:

$$w_{\text{int}} = \frac{W_{\text{int}}}{V} = \frac{1}{V}\int_0^u F\, d\hat{u} = \int_0^u \frac{F\, d\hat{u}}{A\,L} = \int_0^\varepsilon \sigma\, d\hat{\varepsilon}\,. \tag{5.40}$$

The work of the external force is expressed as $W_{\text{ext}} = -Fu$ and the overall potential or potential energy of a system can be expressed as:

$$W = W_{\text{int}} + W_{\text{ext}}\,, \tag{5.41}$$

from which we can show that $W = -W_{\text{int}}$ holds, see Fig. 5.23. If we rearrange HOOKE's law for a linear-elastic spring, i.e. $F = ku$, in the form $u = \frac{1}{k}F = DF$, where D is the so-called compliance, the overall potential can be written as:

$$W = W_{\text{int}} + W_{\text{ext}} = \frac{1}{2}DF^2 - DF^2 = -\frac{1}{2}DF^2\,. \tag{5.42}$$

For practical calculations in this chapter, Table 5.8 summarizes a few expressions for the elastic strain energy for some simple mechanical members.

Let us consider in the following a CT specimen under mode I loading as schematically shown in Fig. 5.24. The potential energy of this system can be expressed

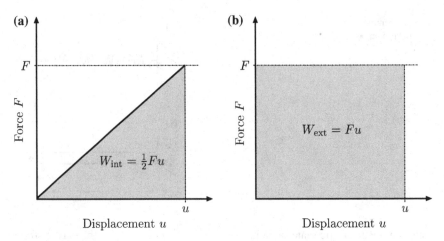

Fig. 5.23 Graphical representation of the **a** elastic strain energy and **b** the work of the external loads

Table 5.8 Elastic strain energies for different simple mechanical members (linear-elastic material assumed)

Member	W_{int}
Bar (tension)	$\int_0^L \dfrac{F^2(x)}{2EA}\,\mathrm{d}x$
Beam (bending)	$\int_0^L \dfrac{M^2(x)}{2EI}\,\mathrm{d}x$
Bar (torsion)	$\int_0^L \dfrac{T^2(x)}{2GI_{\text{p}}}\,\mathrm{d}x$

by Eq. (5.42) where the compliance is now a function of the crack length a, i.e. $D = D(a)$, the material and the entire geometry of the specimen.

Fig. 5.24 CT specimen under mode I loading. Adapted from [25]

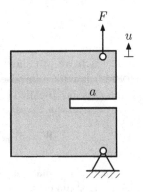

Fig. 5.25 CT specimen under **a** constant load and **b** constant strain condition. Adapted from [1]

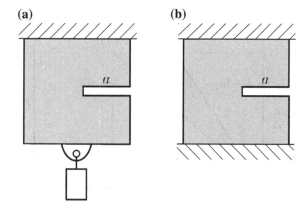

Let us first consider the special case of a constant force F, see Fig. 5.25a, which can be imagined as dead loaded. Then, the potential energy reads:

$$W = -\frac{1}{2}D(a)F^2 \bigg|_{F=\text{const.}} \qquad (5.43)$$

If we consider on the other hand the case of fixed displacement, see Fig. 5.25b, the potential energy reads:

$$W = -\frac{1}{2}\frac{1}{D(a)}u^2 \bigg|_{u=\text{const.}} \qquad (5.44)$$

These energies increase as soon as the crack extends as schematically shown in Fig. 5.26.

Based on an energy approach proposed by GRIFFITH and modified by IRWIN [1], the so-called energy release rate is defined as:

$$\mathcal{G} = -\frac{\mathrm{d}W}{\mathrm{d}\mathcal{A}} \quad \text{or} \quad \mathcal{G} = +\frac{\mathrm{d}W_{\text{int}}}{\mathrm{d}\mathcal{A}}, \qquad (5.45)$$

where $\mathcal{A}$ is the crack area. If this energy release rate reaches a critical value, the so-called critical energy release rate $\mathcal{G}_{\text{c}}$, the crack extends:

$$\mathcal{G} = \mathcal{G}_{\text{c}}. \qquad (5.46)$$

Let us calculate now the energy release rate for the constant force case (assumed thickness: b), see Fig. 5.25a:

$$\mathcal{G} = \frac{\mathrm{d}W_{\text{int}}}{\mathrm{d}\mathcal{A}} = \frac{1}{b}\frac{\mathrm{d}W_{\text{int}}}{\mathrm{d}a} = \frac{1}{b}\frac{\mathrm{d}}{\mathrm{d}a}\left(\frac{1}{2}D(a)F^2\right)_F = \frac{F^2}{2b}\left(\frac{\mathrm{d}D(a)}{\mathrm{d}a}\right)_F. \qquad (5.47)$$

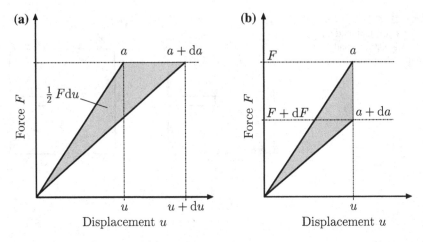

Fig. 5.26 Increase of elastic strain energy due to crack extension: **a** constant force and **b** constant displacement. Adapted from [1]

Based on the same approach, the energy release rate for the constant displacement case (assumed thickness: b), see Fig. 5.25b, reads:

$$\mathcal{G} = \frac{dW_{int}}{d\mathcal{A}} = \frac{1}{b}\frac{dW_{int}}{da} = \frac{1}{b}\frac{d}{da}\left(\frac{1}{2}\frac{1}{D(a)}u^2\right)_u = -\frac{u^2}{2b}\frac{1}{D^2(a)}\left(\frac{dD(a)}{da}\right)_u . \quad (5.48)$$

If we consider in Eq. (5.48) the identity $u = DF$, then it follows that the absolute values of Eqs. (5.48) and (5.47) are the same. The same conclusion can be taken from Fig. 5.26. The difference between the constant force and constant displacement approach is $\frac{1}{2}dFdu$, which is infinitesimal small compared to dW (i.e. the grey shaded area).

5.6 Example: Double cantilever beam (mode I)

Given is a double cantilever beam (DCB) as shown in Fig. 5.27. This configuration is common to measure the fracture toughness of materials. Determine the strain energy release rate.

5.6 Solution

If we consider both arms as cantilever beams of length a, the classical beam theory gives for a single arm the following end deflection:

$$u_1 = \frac{Fa^3}{3EI}. \quad (5.49)$$

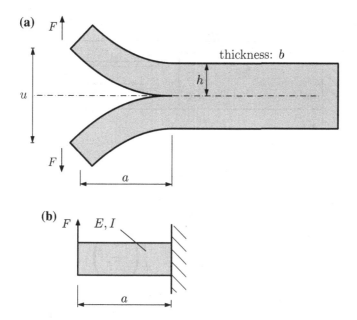

Fig. 5.27 Double cantilever beam under mode I crack opening

Thus, the total crack opening displacement u is obtained as

$$u = \underbrace{\frac{2a^3}{3EI}}_{D} \times F ,$$

(5.50)

from which the compliance D can be identified. The strain energy release rate results from this as:

$$\mathcal{G} = \frac{\mathrm{d}W_{\mathrm{int}}}{\mathrm{d}\mathcal{A}} = \frac{F^2}{2b} \frac{\mathrm{d}}{\mathrm{d}a} \left(\frac{2a^3}{3EI} \right) = \frac{F^2a^2}{bEI} = \frac{12F^2a^2}{b^2Eh^2} .$$

(5.51)

An alternative solution is obtained by considering the elastic strain energies W_{int}, see Table 5.8:

$$W_{\mathrm{int}} = 2 \times \int_0^2 \frac{M^2(x)}{2EI} \mathrm{d}x = \frac{F^2}{EI} \int_0^a (a-x)^2 \mathrm{d}x = \frac{F^2}{EI} (\tfrac{1}{3}-1+1)a^3 = \frac{F^2a^3}{3EI} .$$

(5.52)

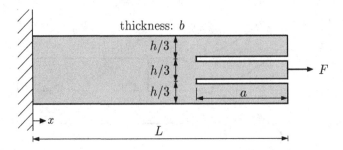

Fig. 5.28 Double cracked bar with a point load

Thus, the strain energy release rate is obtained as:

$$\mathcal{G} = \frac{\mathrm{d}W_{\mathrm{int}}}{\mathrm{d}\mathcal{A}} = \frac{1}{b}\frac{\mathrm{d}W_{\mathrm{int}}}{\mathrm{d}a} = \frac{1}{b}\left(\frac{F^2 a^2}{EI}\right). \tag{5.53}$$

5.7 Example: Double cracked bar with a point load

Given a bar with two parallel cracks of length a, see Fig. 5.28. The middle part is loaded by a single force F. Determine the strain energy release rate of the system.

5.7 Solution

We can split the system in two parts which are effectively loaded. A left-hand part ($0 \le x \le L - a$) of height h and a single right-hand part ($L - a \le x \le L$) of height $\frac{h}{3}$. The elastic strain energy W_{int} of the left-hand part is obtained as:

$$W_{\mathrm{int}} = \int\limits_{0}^{L-a} \frac{F^2}{2EA_1}\mathrm{d}x = \frac{F^2(L-a)}{2EA_1}, \tag{5.54}$$

while the elastic strain energy W_{int} of the right-hand part is:

$$W_{\mathrm{int}} = \int\limits_{0}^{a} \frac{F^2}{2EA_3}\mathrm{d}x = \frac{F^2 a}{2EA_3}. \tag{5.55}$$

Thus, the total elastic strain energy is the sum of both contributions:

$$W_{\mathrm{int}} = \frac{F^2(L-a)}{2EA_1} + \frac{F^2 a}{2EA_3}. \tag{5.56}$$

The strain energy release rate of the system results as:

$$
\begin{aligned}
\mathcal{G} &= \frac{\mathrm{d}W_{\mathrm{int}}}{\mathrm{d}\mathcal{A}} = \frac{1}{b}\frac{\mathrm{d}W_{\mathrm{int}}}{\mathrm{d}(2a)} = \frac{1}{2b}\frac{\mathrm{d}W_{\mathrm{int}}}{\mathrm{d}a} \\
&= \frac{1}{2b}\frac{\mathrm{d}}{\mathrm{d}a}\left(\frac{F^2(L-a)}{2EA_1} + \frac{F^2 a}{2EA_3}\right) \\
&= \frac{1}{2b}\left(-\frac{F^2}{2EA_1} + \frac{F^2}{2EA_3}\right) = \frac{1}{2b}\left(-\frac{F^2}{2Ebh} + \frac{3F^2}{2Ebh}\right) \\
&= \frac{F^2}{2Eb^2h}.
\end{aligned}
\tag{5.57}
$$

It should be noted here that the two fracture parameters, i.e. the stress intensity factor und the strain energy release rate, can be uniquely related for linear elastic material behavior by the following general equation [1]:

$$
\mathcal{G} = \frac{K_{\mathrm{I}}^2}{E'},
\tag{5.58}
$$

where $E' = E$ for plane stress and $E' = \frac{E}{1-\nu^2}$ for plane strain. In the case of simultaneously acting crack opening modes, Eq. (5.58) can be generalized to the following form [1]:

$$
\mathcal{G} = \frac{K_{\mathrm{I}}^2}{E'} + \frac{K_{\mathrm{II}}^2}{E'} + \frac{K_{\mathrm{III}}^2}{2G}.
\tag{5.59}
$$

5.2 Elasto-Plastic Fracture Mechanics

5.2.1 J-Integral

The previous sections introduced the concepts of the stress intensity factor (K) and the strain energy release rate ($\mathcal{G}$). Both concepts are based on the assumption of linear-elastic material behavior or for the case where a very small plastic zone (see Fig. 5.21) is dominated by the surrounding elastic part. The consideration of elasto-plastic material behavior was introduced by RICE [51, 53] who applied the deformation theory of plasticity to derive the mathematical framework for the so-called J-integral. This theory assumes that the elasto-plastic material behavior can be described based on nonlinear elasticity theory as long as monotonic loading occurs, i.e. as long as the body is not unloaded. Thus, the application to cyclic loading conditions is not justified by the applied deformation theory. We will start first with a definition of the J-integral which is similar to the concept of the strain energy release rate ($\mathcal{G}$) of Sect. 5.1.4:

$$J = -\frac{\mathrm{d}W}{\mathrm{d}\mathcal{A}} \quad \text{or} \quad J = +\frac{\mathrm{d}W_{\text{int}}}{\mathrm{d}\mathcal{A}}, \tag{5.60}$$

where $\mathcal{A}$ is the crack area and the material is nonlinear elastic. If this J-integral reaches a critical value, the so-called critical J-integral J_c, the crack extends:

$$J = J_c . \tag{5.61}$$

In the special case of linear-elastic material, the J-integral can be expressed as, see Eqs. (5.45) and (5.58):

$$J^{\text{el}} = \mathcal{G} = \frac{K_{\text{I}}^2}{E'} . \tag{5.62}$$

Let us follow the approach in Sect. 5.1.4 and calculate the J-integral for the constant force and constant displacement case (assumed thickness: b), see Fig. 5.25:

$$J = \frac{\mathrm{d}W_{\text{int}}}{\mathrm{d}\mathcal{A}} = \frac{1}{b}\frac{\mathrm{d}W_{\text{int}}}{\mathrm{d}a}, \tag{5.63}$$

which can be written for the constant force and constant displacement case as [52]:

$$J = \frac{1}{b}\left(\frac{\partial}{\partial a}\int_0^F u\,\mathrm{d}\hat{F}\right)_F = \frac{1}{b}\int_0^F \left(\frac{\partial u}{\partial a}\right)_F \mathrm{d}\hat{F}, \tag{5.64}$$

$$J = -\frac{1}{b}\left(\frac{\partial}{\partial a}\int_0^u F\,\mathrm{d}\hat{u}\right)_u = -\frac{1}{b}\int_0^u \left(\frac{\partial F}{\partial a}\right)_u \mathrm{d}\hat{u}. \tag{5.65}$$

The corresponding strain energies are shown in Fig. 5.29.

A more mathematical definition of the J-integral was given by RICE as a two-dimensional formulation

$$J = \int_\Gamma \left(w_{\text{int}}\mathrm{d}y - T_i\frac{\partial u_i}{\partial x}\mathrm{d}s\right), \tag{5.66}$$

where w_{int} is the volumetric strain energy according to Eq. (5.40), T_i are the components of the traction vector ($T_i = \sigma_{ij}n_i$), u_i are the components of the displacement vector, and $\mathrm{d}s$ is a small line segment along the boundary Γ, see Fig. 5.30.

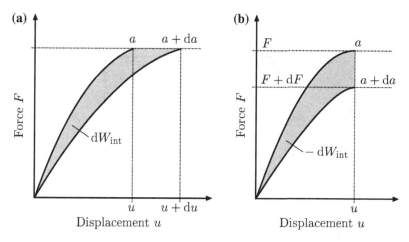

Fig. 5.29 Increase of elastic strain energy due to crack extension for nonlinear material behavior: **a** constant force and **b** constant displacement. Adapted from [1]

Fig. 5.30 Definition of the J-integral as a line integral

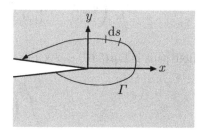

5.8 Example: Double cantilever beam for the evaluation of J[2]

Given is a cracked double cantilever beam as shown in Fig. 5.31. Both legs are loaded by a single moment M_0. The dimensions of the problem can be assumed in such a way that the EULER- BERNOULLI beam theory can be applied. Calculate the value of the J-integral for linear-elastic material behavior.

5.8 Solution

The path for the evaluation of the line integral of Eq. (5.66) is indicated in Fig. 5.31 by the dashed lines and can be split into the five segments $\Gamma_1 \dots \Gamma_5$. It holds along Γ_2 and Γ_4 that $dy = 0$ and $T_i = 0$. Furthermore, it holds along Γ_3 that $w_{int} = 0$ and $T_i = u_i = 0$. Thus, the line integral (5.66) reads as follows:

---

[2]This example is adopted from [60].

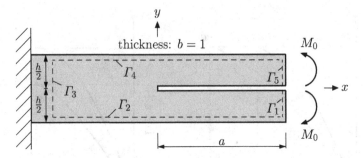

Fig. 5.31 Double cantilever cracked beam under constant bending moment

$$J = \int\limits_{\Gamma_1} \left(w_{\text{int}} dy - T_i \frac{\partial u_i}{\partial x} ds \right) + \int\limits_{\Gamma_5} \left(w_{\text{int}} dy - T_i \frac{\partial u_i}{\partial x} ds \right)$$

$$= \int\limits_{0}^{-h/2} \left(w_{\text{int}} dy - T_i \frac{\partial u_i}{\partial x} ds \right) + \int\limits_{h/2}^{0} \left(w_{\text{int}} dy - T_i \frac{\partial u_i}{\partial x} ds \right). \qquad (5.67)$$

Both parts in Eq. (5.67) can be joined together to obtain the following expression:

$$J = \int\limits_{h/2}^{-h/2} \left(w_{\text{int}} dy - T_i \frac{\partial u_i}{\partial x} ds \right). \qquad (5.68)$$

For a general two-dimensional problem, the relationship between the traction vector T_i and the stress tensor σ_{ij} reads:

$$\begin{bmatrix} T_x \\ T_y \end{bmatrix} = \begin{bmatrix} \sigma_x & \sigma_{xy} \\ \sigma_{xy} & \sigma_y \end{bmatrix} \begin{bmatrix} n_x \\ n_y \end{bmatrix}, \qquad (5.69)$$

where n_i is the unit vector normal to Γ. Applied to our problem of two EULER-BERNOULLI beams, we have $\sigma_{xy} = \sigma_y = 0$. The normal vector to Γ_1 und Γ_5 has the components $[1, 0]^{\text{T}}$ and we can extract from Eq. (5.69) that $T_x = \sigma_x$ holds. Thus, the expression for the J-integral reads now with $ds \rightarrow dy$:

$$J = \int\limits_{h/2}^{-h/2} \left(w_{\text{int}} dy - \sigma_x \frac{\partial u_x}{\partial x} dy \right). \qquad (5.70)$$

Under consideration of $\frac{\partial u_x}{\partial x} = \varepsilon_x$ and $\sigma_x \varepsilon_x = 2w_{\text{int}}$, we can write:

$$J = - \int_{h/2}^{-h/2} w_{\text{int}} \, dy = \int_{-h/2}^{+h/2} w_{\text{int}} \, dy = 2 \int_{0}^{h/2} w_{\text{int}} \, dy . \qquad (5.71)$$

Shifting the origin of the coordinate system to the neutral axis of the upper beam, i.e. $y' = y - h/4$ and $dy' = dy$, gives:

$$J = 2 \int_{-h/4}^{h/4} w_{\text{int}} \, dy' . \qquad (5.72)$$

Considering now the expression for the stress distribution in an EULER- BERNOULLI beam, i.e. $\sigma_x = -\frac{M_z y'}{I_z}$ with $I_z = \frac{1(h/2)^3}{12}$, allows to express the volumetric strain energy as:

$$w_{\text{int}} = \frac{1}{2} \sigma_x \varepsilon_x = \frac{\sigma_x^2}{2E} = \frac{M_0^2 y'^2}{2E I_z^2} . \qquad (5.73)$$

Thus, the J-integral is obtained as:

$$J = 2 \int_{-h/4}^{h/4} \frac{M_0^2 y'^2}{2E I_z^2} \, dy' = \frac{M_0^2}{E I_z^2} \int_{-h/4}^{h/4} y'^2 \, dy' = \frac{M_0^2}{E I_z} . \qquad (5.74)$$

5.2.2 Determination of J_c

The experimental determination of the J-integral uses a similar approach as the additive composition of the strains by their elastic and plastic parts as given in Eq. 3.1, i.e. the additive composition of its elastic and plastic part:

$$J = J^{\text{el}} + J^{\text{pl}} . \qquad (5.75)$$

Considering in Eq. (5.58) the plane strain formulation, we can write that

$$J = \frac{K_I^2}{E}(1 - \nu^2) + \frac{\eta W_{\text{tot}}}{b(w - a)} , \qquad (5.76)$$

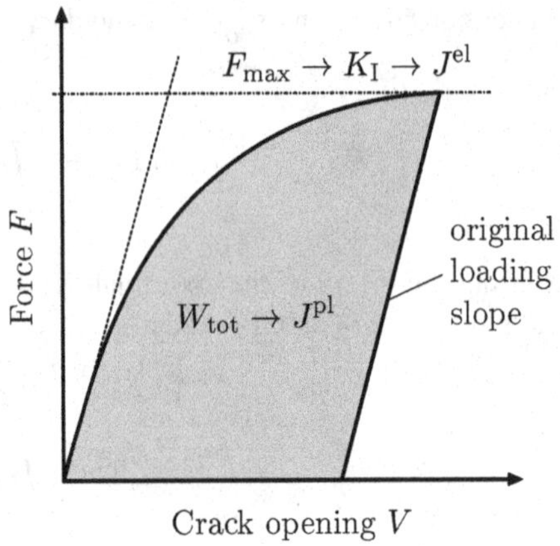

Fig. 5.32 Representation of
the area under the load
versus crack opening curve
according to [4]

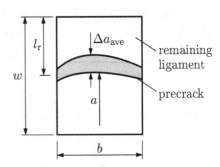

Fig. 5.33 Definition of the
average crack extension
Δa_{ave}. Crack length a
according to Fig. 5.16

where

$$\eta = \begin{cases} 2 & \text{for SE (B), see Table 5.3} \\ 2 + 0.522 \times \frac{w-a}{w} & \text{for C (T), see Table 5.4} \end{cases} . \qquad (5.77)$$

The quantity W_{tot} in Eq. (5.76) refers to the area under the force versus crack opening curve (see Fig. 5.32) while $w - a$ is the remaining ligament (see Fig. 5.33). Further details on the experimental approaches in fracture mechanics are given in the review article [70].

Let us review at the end of this section the introduced failure theories, see Fig. 5.34. Classical plasticity theory considers the local stresses and judges based on a yield condition. In the case of cracks, local stresses cannot be applied and other concepts such as K, $\mathcal{G}$ or J must be applied.

Fig. 5.34 Different failure criteria: **a** stress concentration and **b** stress intensity

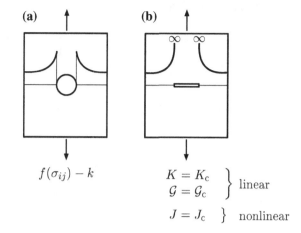

(a)

(b)

$$f(\sigma_{ij}) - k$$

$$\left. \begin{array}{l} K = K_c \\ \mathcal{G} = \mathcal{G}_c \end{array} \right\} \text{ linear}$$

$$J = J_c \quad \} \quad \text{nonlinear}$$

5.3 Supplementary Problems

5.9 Knowledge questions

- State the value of the stress concentration factor for a circular hole in an infinite plate for the case of uniaxial loading of the outer plate edges.
- State the effect of the finite width of a semi-infinite plate with circular hole on the stress concentration factor.
- What aspects does the stress concentration factor depend on?
- Explain the difference between a stress concentration and a stress singularity.
- What aspects does the stress intensity factor depend on?
- Describe some common specimen types to experimentally determine the fracture toughness.
- State a major difference between the stress intensity factor K and the strain energy release rate $\mathcal{G}$.
- State a major difference between the J-integral on the one hand and the stress intensity factor or the strain energy release rate on the other hand.

5.10 Stress state on the edge of a circular hole
Consider an infinite plate as shown in Fig. 5.1 and find the most critical location on the edge of the hole in regards to the acting stresses. Consider the range $0 \leq \varphi \leq 90°$ for $r = a$.

5.11 Critical stress state in an infinite plate with a circular hole
Consider an infinite plate as shown in Fig. 5.1 and calculate the stress states for $\varphi = 30°, 45°, 90°$ and $r = a, 1.25a, 1.5a, 1.75a, 2.0a, 2.25a$. Assume that the external load is equal to $\sigma = 120$ MPa and that the hole's diameter is equal to $2a = 20$ mm. Calculate the stress components σ_r, σ_φ and $\tau_{r\varphi}$ and based on these values the effective VON MISES stress σ_{eff}. Identify the most critical region in regards to the stress state.

Fig. 5.35 Infinite plate with
two holes

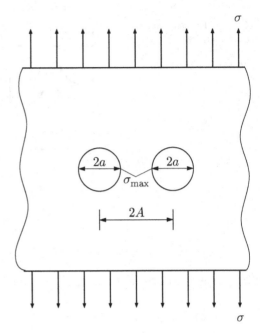

5.12   Critical stress state in an infinite plate with two circular holes

Consider an infinite plate with two holes as shown in Fig. 5.35. The stress concentration factor K_t can be taken from reference works (e.g. [45]) as follows:

$$K_t = \frac{\sigma_{max}}{\sigma_{nom}} = \frac{\sigma_{max}}{\frac{\sigma\sqrt{1-(a/A)^2}}{1-a/A}} = \tag{5.78}$$

$$= 3.0 - 3.0018\left(\frac{a}{A}\right)^1 + 1.0099\left(\frac{a}{A}\right)^2, \tag{5.79}$$

which is valid in the range of $0 \le \frac{a}{A} \le 1$. Investigate the influence on the maximum stress if two holes are located closer and closer.

5.13   Semi-infinite plate with a hole and displacement boundary condition — elasto-plastic material behavior

Given is a thin plate as shown in Fig. 5.36. The geometrical dimensions are $L = 250$ mm, $2A = 100$ mm, $2a = 20$ mm and $t = 10$ mm. The plate is elongated by a constant displacement of $2u = 0.555$ mm. Assume three different types of engineering metals, i.e. the aluminium alloy 7075 ($E_{Al} = 71000$ MPa, $k_{Al}^{init} = 495$ MPa), the titanium alloy Ti-6Al-4V ($E_{Ti} = 110000$ MPa, $k_{Ti}^{init} = 910$ MPa) and the steel alloy 4340 ($E_{St} = 207000$ MPa, $k_{St}^{init} = 1420$ MPa), and check if any location of the plate undergoes plastic deformation. To account for simplifications and assumptions, a safety factor of 2 should be applied.

Fig. 5.36 Semi-infinite plate
with a circular hole and
displacement boundary
condition

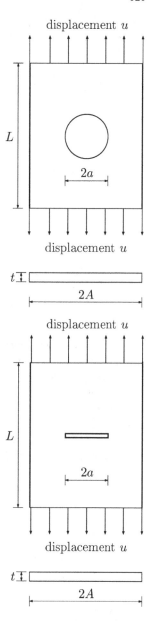

displacement u

L

$2a$

displacement u

t

$2A$

Fig. 5.37 Semi-infinite plate
with a crack and
displacement boundary
condition

displacement u

L

$2a$

displacement u

t

$2A$

5.14 Semi-infinite plate with a crack and displacement boundary condition
Given is a thin plate as shown in Fig. 5.37. The geometrical dimensions are $L = 250$ mm, $2A = 100$ mm, $2a = 20$ mm and $t = 10$ mm. The plate is elongated by a constant displacement of $2u = 0.3$ mm. Assume three different types of engineering metals, i.e. the aluminium alloy 7075 ($E_{Al} = 71000$ MPa, $k_{Al}^{init} = 495$ MPa, $K_{Ic,Al} = 24$ MPa$\sqrt{m}$), the titanium alloy Ti-6Al-4V ($E_{Ti} = 110000$ MPa,

Fig. 5.38 Crack propagating
from a circular hole

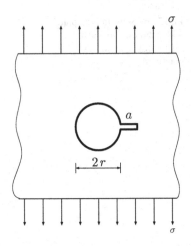

$k_{\mathrm{Ti}}^{\mathrm{init}} = 910$ MPa, $K_{\mathrm{Ic,Ti}} = 55$ MPa$\sqrt{\mathrm{m}}$) and the steel alloy 4340 ($E_{\mathrm{St}} = 207000$ MPa, $k_{\mathrm{St}}^{\mathrm{init}} = 1420$ MPa, $K_{\mathrm{Ic,St}} = 87.4$ MPa$\sqrt{\mathrm{m}}$), and check if the plate would fail under the given load. To account for simplifications and assumptions, a safety factor of 2 should be applied.

5.15 Crack propagating from a circular hole

Given is an infinite thin plate as shown in Fig. 5.38 where a small crack is emanating from a circular hole. This situation is common in engineering practice, e.g. in aircraft structures which contain rivet holes, since cracks emanate from regions of high stress concentration. The geometrical dimensions are $r = 5$ mm and $a = r/10$ and the material is the aluminium alloy 7075 ($k_{\mathrm{Al}}^{\mathrm{init}} = 495$ MPa, $K_{\mathrm{Ic,Al}} = 24$ MPa$\sqrt{\mathrm{m}}$). The plate is loaded by a boundary edge load of $\sigma = 150$ MPa. Under the assumption that $a \ll r$, the stress intensity factor can be expressed as [28]:

$$K_{\mathrm{I}} \approx 1.12(3\sigma)\sqrt{\pi a} \quad \text{for} \quad a \ll r. \tag{5.80}$$

Check for failure and consider in addition the case of the hole without any crack and the case of the crack without hole.

5.16 Crack in a hydraulic cylinder

Given is a hydraulic cylinder[3] as schematically shown in Fig. 5.39. The diameter is $d = 8$ cm and the thickness is $t = 1$ cm. Safety regulations demand a single fluid overpressurization check which generates a hoop stress of 50 % $k_{\mathrm{t}}^{\mathrm{init}}$. In addition, the cylinder design assumes an operating internal fluid pressure corresponding to a hoop stress of 25 % $k_{\mathrm{t}}^{\mathrm{init}}$. The cylinder is made of the aluminium alloy A 7049-T73 ($k_{\mathrm{t}}^{\mathrm{init}} = 460$ MPa and $K_{\mathrm{Ic}} = 23$ MPa$\sqrt{\mathrm{m}}$). The quality insurance team discovered before the fluid overpressurization check a semicircular surface crack of 2 mm depth. The orientation of the crack is normal to the hoop stress direction. Investigate if the

---

[3]This example is adapted from [28].

Fig. 5.39 Crack in a
hydraulic cylinder ('L'
longitudinal and 'H' hoop
direction)

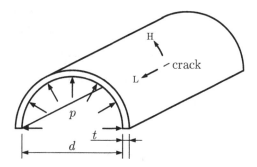

cylinder survives the overpressurization check and if the cylinder experiences a leak-before-break condition during normal operation. The stress intensity factor for the surface crack (see Fig. 5.40) is given as [28, 41]

$$K_{\mathrm{I}} = \sigma\sqrt{\frac{\pi a}{Q}} \times Y_{\mathrm{c}}, \qquad (5.81)$$

where Q is the shape factor for elliptical cracks as:

$$Q = 1 + 1.464\left(\frac{a}{c}\right)^{1.65}. \qquad (5.82)$$

5.17 Calibration of an extensometer

Consider an extensometer which is schematically shown in Fig. 5.41. The measuring principle is based on the deformation of a beam to which a strain gauge is attached. Derive a calibration curve which relates the strain of the specimen and the strain

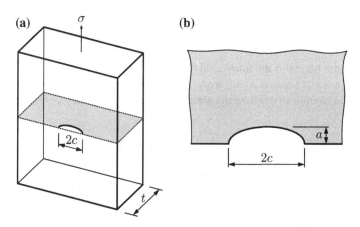

Fig. 5.40 Surface crack in a finite plate: **a** general configuration and **b** elliptical crack geometry

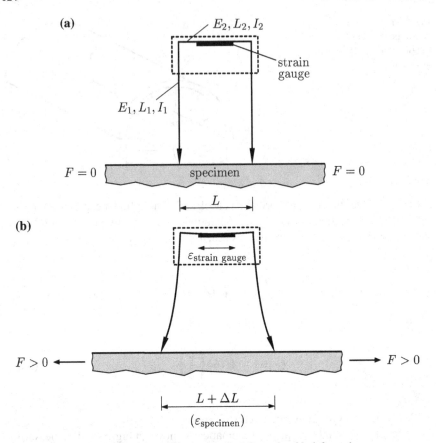

Fig. 5.41 Schematic sketch of an extensometer: **a** undeformed and **b** deformed

measured by the strain gauge ($i = 1, 2$):

$$\varepsilon_{\text{specimen}} = \varepsilon_{\text{specimen}}(\varepsilon_{\text{strain gauge}}, E_i, L_i, I_i, L, h), \qquad (5.83)$$

where h is the height of the horizontal beam ('2'). Assume for simplicity two ideal
EULER- BERNOULLI beams which are treated separately. Depending on the deformed
shape of the beams, an engineering assumption for the boundary conditions should
be taken.

5.18 Difference between classical plasticity theory and linear-elastic fracture mechanics

Consider the case of classical plasticity theory and linear-elastic fracture mechanics
which are applied in the case of a plate with a hole and a plate with a crack, see
Fig. 5.42. Describe the major difference in regards to the considered stress to evaluate
the failure criteria in the framework of both theories.

Fig. 5.42 Different
theoretical approaches:
a classical plasticity theory
and **b** linear-elastic fracture
mechanics

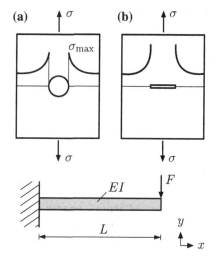

Fig. 5.43 Cantilevered
beam under end load

5.19 Elastic strain energy of a uniform bar

Show that the elastic strain energy of a uniform bar which is loaded by a single end
load F can be expressed for linear-elastic material behavior as:

$$W_{\text{int}} = \int\limits_0^u F\,\mathrm{d}\hat{u} = \int\limits_0^L \frac{F^2}{2EA}\,\mathrm{d}x\,. \tag{5.84}$$

5.20 Elastic strain energy—Castigliano's theorem

Given is an EULER- BERNOULLI beam as shown in Fig. 5.43. Apply CASTIGLIANO's
theorem to calculate the displacement of the point of load application.

5.21 Double cracked bar—energy release rate

Consider a cracked bar of length L which is loaded by three single forces, see
Fig. 5.44. Calculate the strain energy release rate $\mathcal{G}_{\text{II}}$ based on the total strain energy.

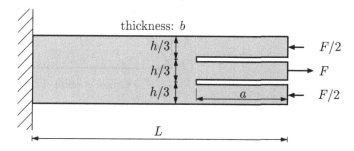

Fig. 5.44 Cracked bar with three point loads

Fig. 5.45 Cracked beam
with point loads

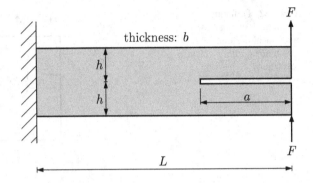

5.22 Cracked beam—energy release rate

Consider a cracked beam of length L which is loaded by two parallel single forces F,
see Fig. 5.45. Calculate the strain energy release rate $\mathcal{G}_{\mathrm{II}}$ based on the total strain
energy.

Fig. 5.46 **a** Geometry of
standard bend specimen and
b force versus crack opening
diagram

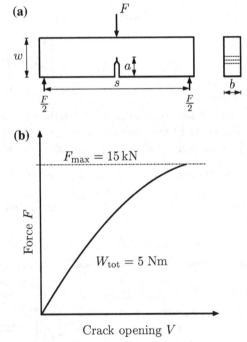

5.23 Standard bend specimen—evaluation of J-integral[4]

Consider a standard bend specimen (SE (B)) loaded in three-point bending, see Fig. 5.46a. The geometrical dimensions of the specimen are $b = 10$ mm, $w = 20$ mm, $s = 80$ mm, and $a = 10$ mm and the elastic material properties of the metal alloy are $E = 205000$ MPa and $\nu = 0.3$.

The obtained force versus crack opening curve is shown in Fig. 5.46b. Evaluate the J-integral for this material based on its elastic and plastic fraction.

---

[4]This example is adapted from [28].

Appendix A
Mathematics

A.1 Greek Alphabet

See Table A.1.

A.2 Frequently Used Constants

$$\pi = 3.14159\,,$$
$$e = 2.71828\,,$$
$$\sqrt{2} = 1.41421\,,$$
$$\sqrt{3} = 1.73205\,,$$
$$\sqrt{5} = 2.23606\,,$$
$$\sqrt{e} = 1.64872\,,$$
$$\sqrt{\pi} = 1.77245\,.$$

A.3 Special Products

$$(x + y)^2 = x^2 + 2xy + y^2\,, \tag{A.1}$$
$$(x - y)^2 = x^2 - 2xy + y^2\,, \tag{A.2}$$
$$(x + y)^3 = x^3 + 3x^2y + 3xy^2 + y^3\,, \tag{A.3}$$
$$(x - y)^3 = x^3 - 3x^2y + 3xy^2 - y^3\,, \tag{A.4}$$
$$(x + y)^4 = x^4 + 4x^3y + 6x^2y^2 + 4xy^3 + y^4\,, \tag{A.5}$$
$$(x - y)^4 = x^4 - 4x^3y + 6x^2y^2 - 4xy^3 + y^4\,. \tag{A.6}$$

© Springer Science+Business Media Singapore 2016
A. Öchsner, *Continuum Damage and Fracture Mechanics*,
DOI 10.1007/978-981-287-865-6

Table A.1 The Greek alphabet

Name	Small letters	Capital letters
Alpha	α	A
Beta	β	B
Gamma	γ	Γ
Delta	δ	Δ
Epsilon	ϵ	E
Zeta	ζ	Z
Eta	η	H
Theta	θ, ϑ	Θ
Iota	ι	I
Kappa	κ	K
Lambda	λ	Λ
Mu	μ	M
Nu	ν	N
Xi	ξ	Ξ
Omicron	o	O
Pi	π	Π
Rho	ρ, ϱ	P
Sigma	σ	Σ
Tau	τ	T
Ypsilon	υ	Υ
Phi	ϕ, φ	Φ
Chi	χ	X
Psi	ψ	Ψ
Omega	ω	Ω

A.4 Trigonometric Functions

Definition at a right-angled triangle

The triangle ABC is in C right-angled and has edges of length a, b, c. The trigonometric functions of the angle α are defined in the following manner (Fig. A.1):

$$\text{sine of } \alpha = \sin \alpha = \frac{a}{c} = \frac{\text{opposite}}{\text{hypotenuse}}, \tag{A.7}$$

$$\text{cosine of } \alpha = \cos \alpha = \frac{b}{c} = \frac{\text{adjacent}}{\text{hypotenuse}}, \tag{A.8}$$

$$\text{tangent of } \alpha = \tan \alpha = \frac{a}{b} = \frac{\text{opposite}}{\text{adjacent}}, \tag{A.9}$$

Fig. A.1 Right-angled
triangle

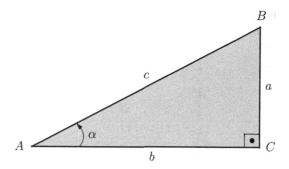

$$\text{cotangent of } \alpha = \cot \alpha = \frac{b}{a} = \frac{\text{adjacent}}{\text{opposite}}, \quad (A.10)$$

$$\text{secant of } \alpha = \sec \alpha = \frac{c}{b} = \frac{\text{hypotenuse}}{\text{adjacent}}, \quad (A.11)$$

$$\text{cosecant of } \alpha = \csc \alpha = \frac{c}{a} = \frac{\text{hypotenuse}}{\text{opposite}}. \quad (A.12)$$

Addition formulae

$$\sin(\alpha \pm \beta) = \sin \alpha \cos \beta \pm \cos \alpha \sin \beta, \quad (A.13)$$

$$\cos(\alpha \pm \beta) = \cos \alpha \cos \beta \mp \sin \alpha \sin \beta, \quad (A.14)$$

$$\tan(\alpha \pm \beta) = \frac{\tan \alpha \pm \tan \beta}{1 \mp \tan \alpha \tan \beta}, \quad (A.15)$$

$$\cot(\alpha \pm \beta) = \frac{\cot \alpha \cot \beta \mp 1}{\cot \beta \pm \cot \beta}. \quad (A.16)$$

Identity formula

$$\sin^2 \alpha + \cos^2 \alpha = 1. \quad (A.17)$$

Analytic values for different angles
Recursion formulae
   See Tables A.2 and A.3.

Table A.2 Analytic values of sine, cosine, tangent and cotangent for different angles

α in degree	α in radian	$\sin\alpha$	$\cos\alpha$	$\tan\alpha$	$\cot\alpha$
0°	0	0	1	0	$\pm\infty$
30°	$\frac{1}{6}\pi$	$\frac{1}{2}$	$\frac{\sqrt{3}}{2}$	$\frac{\sqrt{3}}{3}$	$\sqrt{3}$
45°	$\frac{1}{4}\pi$	$\frac{\sqrt{2}}{2}$	$\frac{\sqrt{2}}{2}$	1	1
60°	$\frac{1}{3}\pi$	$\frac{\sqrt{3}}{2}$	$\frac{1}{2}$	$\sqrt{3}$	$\frac{\sqrt{3}}{3}$
90°	$\frac{1}{2}\pi$	1	0	$\pm\infty$	0
120°	$\frac{2}{3}\pi$	$\frac{\sqrt{3}}{2}$	$-\frac{1}{2}$	$-\sqrt{3}$	$-\frac{\sqrt{3}}{3}$
135°	$\frac{3}{4}\pi$	$\frac{\sqrt{2}}{2}$	$-\frac{\sqrt{2}}{2}$	1	1
150°	$\frac{5}{6}\pi$	$\frac{1}{2}$	$-\frac{\sqrt{3}}{2}$	$-\frac{\sqrt{3}}{3}$	$-\sqrt{3}$
180°	π	0	-1	0	$\pm\infty$
210°	$\frac{7}{6}\pi$	$-\frac{1}{2}$	$-\frac{\sqrt{3}}{2}$	$\frac{\sqrt{3}}{3}$	$\sqrt{3}$
225°	$\frac{5}{4}\pi$	$-\frac{\sqrt{2}}{2}$	$-\frac{\sqrt{2}}{2}$	1	1
240°	$\frac{4}{3}\pi$	$-\frac{\sqrt{3}}{2}$	$-\frac{1}{2}$	$\sqrt{3}$	$\frac{\sqrt{3}}{3}$
270°	$\frac{3}{2}\pi$	-1	0	$\pm\infty$	0
300°	$\frac{5}{3}\pi$	$-\frac{\sqrt{3}}{2}$	$\frac{1}{2}$	$-\sqrt{3}$	$-\frac{\sqrt{3}}{3}$
315°	$\frac{7}{4}\pi$	$-\frac{\sqrt{2}}{2}$	$\frac{\sqrt{2}}{2}$	-1	-1
330°	$\frac{11}{6}\pi$	$-\frac{1}{2}$	$\frac{\sqrt{3}}{2}$	$-\frac{\sqrt{3}}{3}$	$-\sqrt{3}$
360°	2π	0	1	0	$\pm\infty$

Table A.3 Recursion formulae for trigonometric functions

	$-\alpha$	$90°\pm\alpha$	$180°\pm\alpha$	$270°\pm\alpha$	$k(360°)\pm\alpha$
		$\frac{\pi}{2}\pm\alpha$	$\pi\pm\alpha$	$\frac{3\pi}{2}\pm\alpha$	$2k\pi\pm\alpha$
sin	$-\sin\alpha$	$\cos\alpha$	$\mp\sin\alpha$	$-\cos\alpha$	$\pm\sin\alpha$
cos	$\cos\alpha$	$\mp\sin\alpha$	$-\cos\alpha$	$\pm\sin\alpha$	$\cos\alpha$
tan	$-\tan\alpha$	$\mp\cot\alpha$	$\pm\tan\alpha$	$\mp\cot\alpha$	$\pm\tan\alpha$
csc	$-\csc\alpha$	$\sec\alpha$	$\mp\csc\alpha$	$-\sec\alpha$	$\pm\csc\alpha$
sec	$\sec\alpha$	$\mp\csc\alpha$	$-\sec\alpha$	$\pm\csc\alpha$	$\sec\alpha$
cot	$-\cot\alpha$	$\mp\tan\alpha$	$\pm\cot\alpha$	$\mp\tan\alpha$	$\pm\cot\alpha$

A.5 Derivatives

- $\dfrac{d}{dx}\left(\dfrac{1}{x}\right) = -\dfrac{1}{x^2}$

- $\dfrac{d}{dx}x^n = n \times x^{n-1}$

- $\dfrac{d}{dx}\sqrt[n]{x} = \dfrac{1}{n \times \sqrt[n]{x^{n-1}}}$

- $\dfrac{d}{dx}\sin(x) = \cos(x)$

- $\dfrac{d}{dx}\cos(x) = -\sin(x)$

- $\dfrac{d}{dx}\ln(x) = \dfrac{1}{x}$

- $\dfrac{d}{dx}|x| = \begin{cases} -1 \text{ for } x < 0 \\ 1 \text{ for } x > 0 \end{cases}$

- $\dfrac{d}{dx}(f(x) \times g(x)) = \dfrac{df(x)}{dx}g(x) + f(x)\dfrac{dg(x)}{dx}$ (product rule)

- $\dfrac{d}{dx}\left(\dfrac{f(x)}{g(x)}\right) = \dfrac{df(x)/dx \times g(x) - f(x) \times dg(x)/dx}{[g(x)]^2}$ (quotient rule)

A.6 Integrals

The indefinite integral or antiderivative $F(x) = \int f(x)dx + c$ of a function $f(x)$ is a differentiable function $F(x)$ whose derivative is equal to $f(x)$, i.e., $\frac{dF(x)}{dx} = f(x)$. The definite integral of a continuous real-valued function $f(x)$ on a closed interval $[a, b]$, i.e., $\int_a^b f(d)dx = F(b) - F(a)$, is represented by the area under the curve $f(x)$ from $x = a$ to $x = b$.

Some selected antiderivatives (c: arbitrary constant of integration):

- $\int e^x dx = e^x + c$
- $\int \sqrt{x}dx = \frac{2}{3}x^{\frac{3}{2}} + c$
- $\int \sin(x)dx = -\cos(x) + c$
- $\int \cos(x)dx = \sin(x) + c$
- $\int \sin(\alpha x)\cos(\alpha x)dx = \dfrac{1}{2}\sin^2(x) + c$
- $\int \sin^2(\alpha x)dx = \dfrac{1}{2}(x - \sin(x)\cos(x)) + c = \dfrac{1}{2}(x - \frac{1}{2}\sin(2x)) + c$
- $\int \cos^2(\alpha x)dx = \dfrac{1}{2}(x + \sin(x)\cos(x)) + c = \dfrac{1}{2}(x + \frac{1}{2}\sin(2x)) + c$

A.1 Example: Indefinite and definite integral

Calculate the indefinite and definite integral of $f(x) = x^2 + 1$. The definite integral is to be calculated in the interval $[1, 2]$.

A.1 Solution

Indefinite Integral:

$$F(x) = \int (x^2 + 1)dx = \frac{x^3}{3} + x + c. \tag{A.18}$$

Definite Integral:

$$\int_1^2 (x^2 + 1)dx = \left[\frac{x^3}{3} + x \right]_1^2 = \frac{10}{3}. \tag{A.19}$$

A.7 Matrix Operations

A matrix A is defined as a set of quantities a_{ij} ordered as follows

$$A = \begin{array}{c} \xrightarrow{\quad} j \\ \downarrow \\ i \end{array} \begin{bmatrix} a_{11} & a_{12} & \cdots & a_{1n} \\ a_{21} & a_{22} & \cdots & a_{2n} \\ \vdots & \vdots & \ddots & \vdots \\ a_{m1} & a_{m2} & \cdots & a_{mn} \end{bmatrix} \tag{A.20}$$

and denoted by a bold upper case letter. As indicated, the position of a term a_{ij} is defined by indices i and j: the first index determines the row and the second index determines the column in the matrix. For example, a_{23} is the term in the third row and in the second column. The matrix (A.20) is of order $m \times n$. A matrix is said to be square if $m = n$, and rectangular if $m \neq n$. An $m \times 1$ matrix is called a column matrix and denoted by a bold lower case variable. For example, the column matrix b

$$b = \begin{bmatrix} b_1 \\ b_2 \\ \vdots \\ b_m \end{bmatrix} \tag{A.21}$$

is defined as the matrix of order $m \times 1$.

The transposed matrix of a matrix A is denoted A^T and represents the matrix with the columns and rows interchanged, i.e.

$$(A^T)_{ij} = (A)_{ji}. \tag{A.22}$$

If we use symmetric matrices for which the elements are symmetric with respect to the main diagonal, then

$$A^T = A . \tag{A.23}$$

The transposed of the matrix (A.21) results in a row matrix:

$$b^T = \begin{bmatrix} b_1 & b_2 & \dots & b_m \end{bmatrix} . \tag{A.24}$$

The inverse of a matrix can be calculated as follows:

- Equation for a 2×2 matrix:

$$A^{-1} = \begin{bmatrix} a & b \\ c & d \end{bmatrix}^{-1} = \frac{1}{ad - bc} \times \begin{bmatrix} d & -b \\ -c & a \end{bmatrix} . \tag{A.25}$$

- Equation for a 3×3 matrix:

$$A^{-1} = \begin{bmatrix} a & b & c \\ d & e & f \\ g & h & i \end{bmatrix}^{-1} = \frac{1}{\det(A)} \begin{bmatrix} ei - fh & ch - bi & bf - ce \\ fg - di & ai - cg & cd - af \\ dh - eg & bg - ah & ae - bd \end{bmatrix} , \tag{A.26}$$

where

$$\det(A) = aei + bfg + cdh - ceg - afh - bdi . \tag{A.27}$$

Appendix B
Mechanics

B.1 Mohr's Circle

Figure B.1 shows MOHR's circle for a plane stress state $(\sigma_x, \sigma_y, \tau_{xy})$. This representation allows to determine the principal stresses σ_1 and σ_2 or the maximum shear stress τ_{max}, see [24].

$$\sigma_{1,2} = \frac{\sigma_x + \sigma_y}{2} \pm \sqrt{\left(\frac{\sigma_x - \sigma_y}{2}\right)^2 + \tau_{xy}^2}, \tag{B.1}$$

$$\tau_{max} = \sqrt{\left(\frac{\sigma_x - \sigma_y}{2}\right)^2 + \tau_{xy}^2}. \tag{B.2}$$

B.2 Deformation Energy

The expression for the total deformation energy[1] w in the linear-elastic range of a material is derived in the following. Let us consider first a simple tensile test as shown in Fig. B.2 where a normal stress σ is acting on a volume element.

The volumetric deformation energy per unit volume is obtained by:

$$w^\circ = \frac{W^\circ}{V} = \int \frac{F\,\mathrm{d}L}{A\,L} = \int \sigma \mathrm{d}\varepsilon = \frac{1}{2}\sigma\varepsilon. \tag{B.3}$$

---

[1] It should be noted here that this is the deformation energy per unit volume: $w = \frac{W}{V}$.

© Springer Science+Business Media Singapore 2016
A. Öchsner, *Continuum Damage and Fracture Mechanics*,
DOI 10.1007/978-981-287-865-6

137

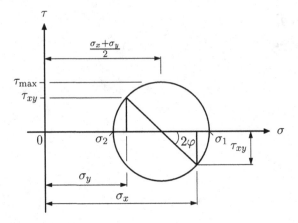

Fig. B.1 Mohr's circle for a plane stress state

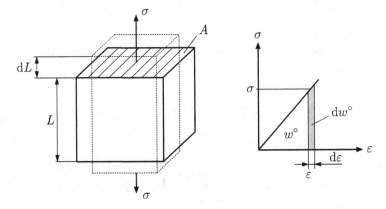

Fig. B.2 Derivation of the volumetric deformation energy

A pure shear stress state as shown in Fig. B.3 is characterized by a shear stress τ acting on a volume element.

The distortional deformation energy per unit volume is obtained by:

$$w^{s} = \frac{W^{s}}{V} = \int \frac{F \, ds}{A \, L} = \int \tau d\gamma = \frac{1}{2}\tau\gamma. \qquad (B.4)$$

In a general three-dimensional stress state, the single components must be superimposed to obtain the total deformation energy per unit volume as:

$$w = \frac{1}{2}\left(\sigma_{xx}\varepsilon_{xx} + \sigma_{yy}\varepsilon_{yy} + \sigma_{zz}\varepsilon_{zz} + \tau_{xy}\gamma_{xy} + \tau_{xz}\gamma_{xz} + \tau_{yz}\gamma_{yz}\right). \qquad (B.5)$$

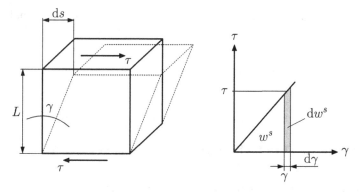

Fig. B.3 Derivation of the distortional deformation energy

Using HOOKE's law according to Eqs. (2.16) and (2.19) to express the strains in terms of stresses, i.e.

$$\varepsilon_{xx} = \frac{1}{E}\left(\sigma_{xx} - \nu(\sigma_{yy} + \sigma_{zz})\right), \tag{B.6}$$

$$\varepsilon_{yy} = \frac{1}{E}\left(\sigma_{yy} - \nu(\sigma_{xx} + \sigma_{zz})\right), \tag{B.7}$$

$$\varepsilon_{zz} = \frac{1}{E}\left(\sigma_{zz} - \nu(\sigma_{xx} + \sigma_{yy})\right), \tag{B.8}$$

$$\gamma_{xy} = \frac{1}{G}\tau_{xy}, \ \gamma_{xz} = \frac{1}{G}\tau_{xz}, \ \gamma_{yz} = \frac{1}{G}\tau_{yz}, \tag{B.9}$$

the total deformation energy can be expressed as:

$$w = \frac{1}{2E}\left(\sigma_{xx}\left[\sigma_{xx} - \nu(\sigma_{yy} + \sigma_{zz})\right] + \sigma_{yy}\left[\sigma_{yy} - \nu(\sigma_{xx} + \sigma_{zz})\right] +$$

$$+ \sigma_{zz}\left[\sigma_{zz} - \nu(\sigma_{xx} + \sigma_{yy})\right]\right) + \frac{1}{2G}\left(\tau_{xy}^2 + \tau_{xz}^2 + \tau_{yz}^2\right). \tag{B.10}$$

Expressing the shear modulus G in terms of E and ν (see Table 2.2), i.e. $G = \frac{E}{2(1+\nu)}$, and doing some mathematical transformations, the total deformation energy per unit volume is finally obtained as:

$$w = \underbrace{\frac{1 - 2\nu}{6E}\left(\sigma_{xx} + \sigma_{yy} + \sigma_{zz}\right)^2}_{w^{\circ}} +$$

$$+ \underbrace{\frac{1 + \nu}{6E}\left[(\sigma_{xx} - \sigma_{yy})^2 + (\sigma_{yy} - \sigma_{zz})^2 + (\sigma_{zz} - \sigma_{xx})^2 + 6(\tau_{xy}^2 + \tau_{yz}^2 + \tau_{xz}^2)\right]}_{w^s}.$$

$$\tag{B.11}$$

The definition of the basic invariants (see Table 3.4) allows a shorter formulation as follows:

$$w = \underbrace{\frac{1-2\nu}{6E} J_1^2}_{w^\circ} + \underbrace{\frac{1+\nu}{E} J_2'}_{w^s} = \underbrace{\frac{1-2\nu}{6E} J_1^2}_{w^\circ} + \underbrace{\frac{1}{2G} J_2'}_{w^s}. \tag{B.12}$$

B.3 Stresses in Thin Cylinders and Shells (Pressure Vessels)

The following section summarizes a few basic results in regards to the stress state in *thin* cylinders and shells [49, 64]. Thin in this context means that the ratio between the radial thickness t and the inner diameter d is smaller than one-tenth of the inner diameter:

$$\frac{t}{d} \le \frac{d}{10}. \tag{B.13}$$

Thin cylinders and shells are widely used as pressure vessels used for gas storage in the form of cylindrical or spherical tanks. The stress state in a thin cylinder loaded by an internal pressure p (see Fig. B.4) is approximately described by the hoop or circumferential stress

$$\sigma_H = \frac{pd}{2t}, \tag{B.14}$$

and the longitudinal stress

$$\sigma_L = \frac{pd}{4t}. \tag{B.15}$$

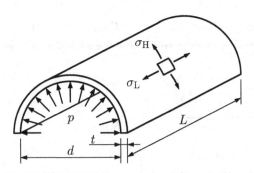

Fig. B.4 Half of a thin cylinder which is loaded by an internal pressure

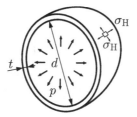

Fig. B.5 Half of a thin spherical shell which is loaded by an internal pressure

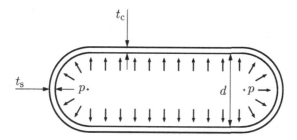

Fig. B.6 Schematic representation of a thin cylinder with hemispherical ends

The stress state in a thin spherical shell loaded by an internal pressure p (see Fig. B.5) is approximately described by two mutually perpendicular circumferential stresses of equal value:

$$\sigma_H = \frac{pd}{4t}. \tag{B.16}$$

The stress state in a thin cylinder with hemispherical ends loaded by an internal pressure p (see Fig. B.6) is approximately described in the cylindrical part by

$$\sigma_{H_c} = \frac{pd}{2t_c}, \quad \sigma_{L_c} = \frac{pd}{4t_c} \tag{B.17}$$

and in the hemispherical ends by

$$\sigma_{H_s} = \frac{pd}{4t_s}. \tag{B.18}$$

B.4 Mixed-Mode: Biaxial Loading

Figure B.7 shows a plate with an inclined crack which is loaded by normal stresses σ_1 and σ_2. The inclination angle φ is defined between the crack and the σ_1-plane. Depending on the stress ratio $b = \frac{\sigma_2}{\sigma_1}$ (biaxiality ratio), the stress intensity factors

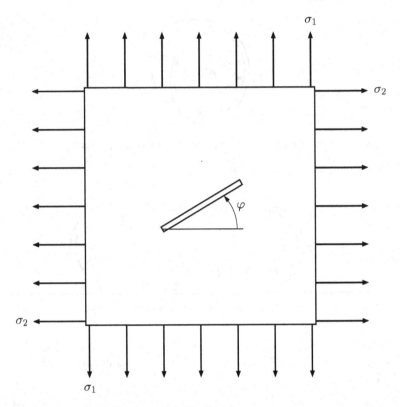

Fig. B.7 Plate under biaxial loading with inclined crack

are given by [1]

$$K_I = K_{I(0)} \left(\cos^2(\varphi) + b \sin^2(\varphi) \right) , \tag{B.19}$$

$$K_{II} = K_{I(0)} \left(\sin(\varphi) \cos(\varphi)(1 - b) \right) , \tag{B.20}$$

where $K_{I(0)}$ is the mode I stress intensity factor for $\varphi = 0$.

The special case for $\varphi = 0$ is obtained as $K_I = K_{I(0)}$ and $K_{II} = 0$, i.e. stresses parallel to the crack can be excluded in the framework of linear-elastic fracture mechanics [34, 50].

Answers to Supplementary Problems

Problems from Chapter 1

1.2 Approximation of the true stress

$$A_0 L_0 = AL \quad \Rightarrow \quad \sigma_{\mathrm{tr}} = \frac{F}{A_0}(1 + \varepsilon) \, . \tag{B.21}$$

1.3 Approximation of the true strain

$$\ln(1 + \varepsilon) = \sum_{n=1}^{\infty} \frac{(-1)^{n+1}}{n} \varepsilon^n = \varepsilon - \frac{\varepsilon^2}{2} + \frac{\varepsilon^3}{4} - \cdots \tag{B.22}$$

Neglect the higher-order expressions for small strain values.

1.4 Engineering versus true stress and strain representation
The stress-strain representations are shown in Fig. B.8.

Problems from Chapter 2

2.2 Approximation of the volume change in a hydrostatic compression test

$$\frac{\Delta V}{V} = \frac{(a + \Delta a)(b + \Delta b)(c + \Delta c) - abc}{abc} \tag{B.23}$$

$$= \frac{a\left(1 + \frac{\Delta a}{a}\right) b\left(1 + \frac{\Delta b}{b}\right) c\left(1 + \frac{\Delta c}{c}\right) - abc}{abc}$$

$$= (1 + \varepsilon_x)(1 + \varepsilon_y)(1 + \varepsilon_z) - 1$$

$$= \varepsilon_x + \varepsilon_y + \varepsilon_z + \underbrace{\varepsilon_x \varepsilon_y \varepsilon_z + \varepsilon_x \varepsilon_y + \varepsilon_x \varepsilon_z + \varepsilon_y \varepsilon_z + 1 - 1}_{\text{small with respect to } \varepsilon_x, \varepsilon_y, \varepsilon_z}$$

$$\approx \varepsilon_x + \varepsilon_y + \varepsilon_z \, . \tag{B.24}$$

2.3 Mohr's circle for simple load cases
MOHR's circle for the different load cases is shown in Fig. B.9.

2.4 Derivation of the evaluation equation for the torsion test
A deformed element of the cylindrical specimen under moment load is shown in Fig. B.10a. For small deformations, the relation between the angles is

$$r \, \mathrm{d}\vartheta = \gamma \mathrm{d}x \, . \tag{B.25}$$

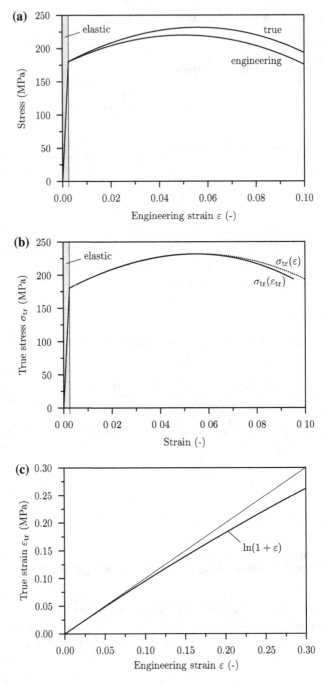

Fig. B.8 **a** Stress versus engineering strain; **b** true stress versus strain; **c** true strain versus engineering strain

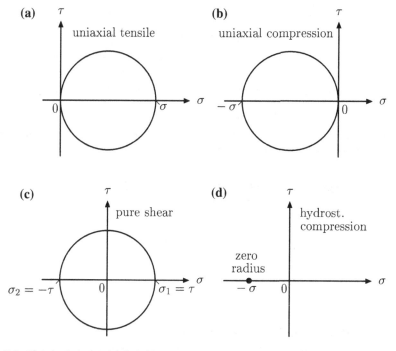

Fig. B.9 Mohr's circle for simple load cases: **a** tensile test, **b** compression test, **c** pure shear test, and **d** hydrostatic compression

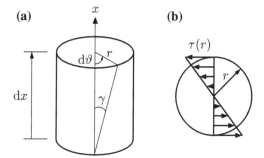

Fig. B.10 Pure shear test: **a** deformed specimen element and **b** shear stress distribution

A combination with HOOKE's law for a pure shear stress state, i.e. $\tau = G\gamma$, gives:

$$\tau = Gr\frac{\mathrm{d}\vartheta}{\mathrm{d}x}. \tag{B.26}$$

Fig. B.11 Realization of a pure shear stress state based on normal stresses

The shear stress distribution shown in Fig. B.10a produces a moment which must balance the external torsional moment M_T:

$$M_T = \int r\tau \, dA \overset{(B.26)}{=} \int r^2 G \frac{d\vartheta}{dx} \, dA = G \frac{d\vartheta}{dx} \underbrace{\int r^2 \, dA}_{I_p} \,. \qquad (B.27)$$

After integration over the twist angle ϑ:

$$M_T L = G I_p \vartheta_L \quad \Leftrightarrow \quad G = \frac{M_T L}{I_p \vartheta_L} \,. \qquad (B.28)$$

2.5 Alternative realization of pure shear teat

Rotation of the coordinate system by 45° gives $\sigma_x = \sigma$ and $\sigma_y = -\sigma$, see Fig. B.11.

2.6 Evaluation of isostatic compression test

$\sigma_{xy} = \sigma_{yz} = \sigma_{xz} = 0$, $\varepsilon_{xy} = \varepsilon_{yz} = \varepsilon_{xz} = 0$. Evaluation of HOOKE's law for the first stress component gives:

$$\sigma_x = K(\varepsilon_x + \varepsilon_y + \varepsilon_z) + G(\tfrac{4}{3}\varepsilon_x - \tfrac{2}{3}\varepsilon_y - \tfrac{2}{3}\varepsilon_z) \,. \qquad (B.29)$$

Doing the same for the other two normal stress components, consideration that the normal strains are the same (in the expression with G) and summing up gives the required relationship.

2.7 Axial compaction

$$\frac{d\sigma_z}{d\varepsilon_z} = K + \frac{4}{3}G = \frac{E}{3(1 - 2\nu)} + \frac{2}{3} \times \frac{E}{1 + \nu} \,. \qquad (B.30)$$

2.8 Plane strain experiment

$$\nu = \frac{\sigma_2}{\sigma_1} \,. \qquad (B.31)$$

2.9 Hooke's law in terms of Lamé's constants

$$
\begin{bmatrix} \sigma_x \\ \sigma_y \\ \sigma_z \\ \sigma_{xy} \\ \sigma_{yz} \\ \sigma_{xz} \end{bmatrix} = \begin{bmatrix} \lambda+2\mu & \lambda & \lambda & 0 & 0 & 0 \\ \lambda & \lambda+2\mu & \lambda & 0 & 0 & 0 \\ \lambda & \lambda & \lambda+2\mu & 0 & 0 & 0 \\ 0 & 0 & 0 & \mu & 0 & 0 \\ 0 & 0 & 0 & 0 & \mu & 0 \\ 0 & 0 & 0 & 0 & 0 & \mu \end{bmatrix} \begin{bmatrix} \varepsilon_x \\ \varepsilon_y \\ \varepsilon_z \\ 2\,\varepsilon_{xy} \\ 2\,\varepsilon_{yz} \\ 2\,\varepsilon_{xz} \end{bmatrix}, \tag{B.32}
$$

or as the elastic compliance form:

$$
\begin{bmatrix} \varepsilon_x \\ \varepsilon_y \\ \varepsilon_z \\ 2\,\varepsilon_{xy} \\ 2\,\varepsilon_{yz} \\ 2\,\varepsilon_{xz} \end{bmatrix} = \begin{bmatrix} \frac{\lambda+\mu}{\mu(3\lambda+2\mu)} & -\frac{\lambda}{2\mu(3\lambda+2\mu)} & -\frac{\lambda}{2\mu(3\lambda+2\mu)} & 0 & 0 & 0 \\ -\frac{\lambda}{2\mu(3\lambda+2\mu)} & \frac{\lambda+\mu}{\mu(3\lambda+2\mu)} & -\frac{\lambda}{2\mu(3\lambda+2\mu)} & 0 & 0 & 0 \\ -\frac{\lambda}{2\mu(3\lambda+2\mu)} & -\frac{\lambda}{2\mu(3\lambda+2\mu)} & \frac{\lambda+\mu}{\mu(3\lambda+2\mu)} & 0 & 0 & 0 \\ 0 & 0 & 0 & \frac{1}{\mu} & 0 & 0 \\ 0 & 0 & 0 & 0 & \frac{1}{\mu} & 0 \\ 0 & 0 & 0 & 0 & 0 & \frac{1}{\mu} \end{bmatrix} \begin{bmatrix} \sigma_x \\ \sigma_y \\ \sigma_z \\ \sigma_{xy} \\ \sigma_{yz} \\ \sigma_{xz} \end{bmatrix}. \tag{B.33}
$$

Problems from Chapter 3

3.2 Evaluation of tensile test with linear hardening

$$
E = 70000 \text{ MPa}, \ E^{\text{pl}} = 7000 \text{ MPa}, \ E^{\text{elpl}} = 6363.636 \text{ MPa}, \tag{B.34}
$$

$$
k(\kappa) = 350 \text{ MPa} + 7000 \text{ MPa} \times \kappa. \tag{B.35}
$$

$$
\sigma(\varepsilon) = E\varepsilon \ \text{(elastic part)}, \tag{B.36}
$$

$$
\sigma(\varepsilon) = (E - E^{\text{elpl}}) \times \varepsilon_{\text{t}}^{\text{init}} + E^{\text{elpl}} \times \varepsilon \ \text{(elasto-plastic part)}. \tag{B.37}
$$

$$
300 \text{ MPa:} \ \varepsilon^{\text{pl}} = 0, \tag{B.38}
$$

$$
600 \text{ MPa:} \ \varepsilon^{\text{pl}} = 0.03571. \tag{B.39}
$$

3.3 Investigation of initial yielding for different materials and boundary conditions

Case (a) The stress is independent of the materials, i.e. $\sigma = F/A$, and this stress must be compared to the three different initial yield stresses.

Case (b) The stress depends on the material, i.e. $\sigma = Eu/L$. These material specific stresses must be compared to the corresponding initial yield stress.

3.4 Calculation of the principal stresses for a plane stress state

• MOHR's circle:

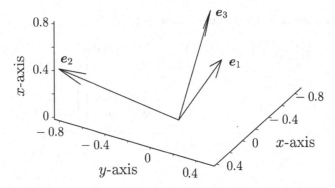

Fig. B.12 Normalized eigenvectors of the stress tensor σ_{ij} in (x, y, z) coordinates

$$\sigma_{1,2} = \frac{\sigma_{xx} + \sigma_{yy}}{2} \pm \sqrt{\left(\frac{\sigma_{xx} - \sigma_{yy}}{2}\right)^2 + \tau_{xy}^2} = 90 \pm 10\sqrt{5}. \qquad (B.40)$$

• tensor approach:

$$\det\left(\begin{bmatrix} 100 - \sigma_i & 20 & 0 \\ 20 & 80 - \sigma_i & 0 \\ 0 & 0 & 0 - \sigma_1 \end{bmatrix}\right) = 0, \qquad (B.41)$$

$$-\sigma_i^3 + 180\sigma_i^2 - 7600\sigma_i = 0, \qquad (B.42)$$

$$\sigma_1 = 90 + 10\sqrt{5} = 112.36, \sigma_2 = 90 - 10\sqrt{5} = 67.64, \sigma_3 = 0. \qquad (B.43)$$

3.5 Calculation of the principal stresses and the corresponding principal directions

Principal stresses: $\sigma_1 = 110$, $\sigma_2 = 70$, $\sigma_3 = 40$.

Normalized principal directions: $e_1 = \frac{1}{\sqrt{14}}[1\ 2\ 3]^T$, $e_2 = \frac{1}{\sqrt{6}}[1\ -2\ 1]^T$, $e_3 = \frac{1}{\sqrt{21}}[-4\ -1\ 2]^T$. The graphical representation is shown in Fig. B.12.

3.6 Calculation of the principal and basic invariants

$$I_1 = 220, I_2 = 14900, I_3 = 308000, \qquad (B.44)$$

$$J_1 = 220, J_2 = 9300, J_3 = 579333.333. \qquad (B.45)$$

3.7 Difference between shear and tensile yield stress for the von Mises and Tresca yield condition

• Tresca:

$$\left(\frac{\sigma}{k_t}\right)^2 + \left(\frac{2\tau}{k_t}\right)^2 - 1 = 0 \ \text{with} \ \sigma \to 0 \Rightarrow \tau \to k_s : k_s = \frac{k_t}{2}. \tag{B.46}$$

- VON MISES:

$$k_s = \frac{k_t}{\sqrt{3}}. \tag{B.47}$$

3.8 Difference between the von Mises and Tresca yield condition in two-component stress spaces

For both cases the same relative difference: 13.4%.

3.9 Hydrostatic stress and Mises/Tresca yield condition

- VON MISES:

$$F = \sqrt{\frac{1}{2}\left[(\sigma_1 - \sigma_2)^2 + (\sigma_2 - \sigma_3)^2 + (\sigma_3 - \sigma_1)^2\right]} - k_t = 0 \tag{B.48}$$

$$\Rightarrow \ 0 - k_t = 0. \tag{B.49}$$

- TRESCA:

$$F = \max\left(\frac{1}{2}|\sigma_1 - \sigma_2|, \frac{1}{2}|\sigma_2 - \sigma_3|, \frac{1}{2}|\sigma_3 - \sigma_1|\right) - k_s = 0 \tag{B.50}$$

$$\Rightarrow \ \max(0, 0, 0) - k_s = 0. \tag{B.51}$$

3.10 Influence of isotropic and kinematic hardening on the shape of the yield surface

$$\left(\frac{\frac{\sigma}{k_t^{\text{init}}}}{1}\right)^2 + \left(\frac{\frac{\tau}{k_t^{\text{init}}}}{\frac{1}{\sqrt{3}}}\right)^2 - 1 \ \text{(initial)}, \tag{B.52}$$

$$\left(\frac{\frac{\sigma}{k_t^{\text{init}}}}{1.2}\right)^2 + \left(\frac{\frac{\tau}{k_t^{\text{init}}}}{\frac{1.2}{\sqrt{3}}}\right)^2 - 1 \ \text{(isotropic)}, \tag{B.53}$$

$$\left(\frac{\sigma}{k_t^{\text{init}}} - 0.2\right)^2 + \left(\frac{1}{\sqrt{3}}\left[\frac{\tau}{k_t^{\text{init}}} - 0.1\right]\right)^2 - 1 \ \text{(kinematic)}. \tag{B.54}$$

The graphical representation of these three equations is shown in Fig. B.13. It can be seen that isotropic hardening results in a uniform expansion of the initial yield surface under maintaining the origin of the ellipse. Kinematic hardening results in a translation of the yield surface while the shape and size remains constant.

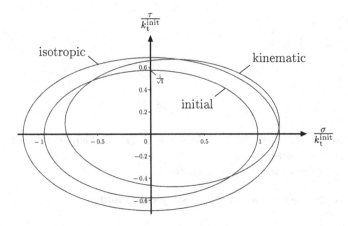

Fig. B.13 Graphical representation of the yield condition according to VON MISES in the σ-τ space

3.11 Thin-walled pressure vessel under internal pressure: von Mises and Tresca yield condition

$$t_{\mathrm{M}} = 6.77\,\mathrm{mm}\,, t_{\mathrm{T}} = 7.81\,\mathrm{mm}\,. \tag{B.55}$$

3.12 Thin-walled pressure vessel under internal pressure and axial force: von Mises and Tresca yield condition

$$t_{\mathrm{M}} = 7.11\,\mathrm{mm}\,, t_{\mathrm{T}} = 7.81\,\mathrm{mm}\,. \tag{B.56}$$

3.13 Thin-walled pressure vessel under internal pressure and twisting moment: von Mises and Tresca yield condition

$$t_{\mathrm{M}} = 7.58\,\mathrm{mm}\,, t_{\mathrm{T}} = 8.63\,\mathrm{mm}\,. \tag{B.57}$$

Problems from Chapter 4

4.3 Uniaxial tensile test with and without consideration of contraction
The following holds for both cases:

$$\varepsilon_x = \frac{\sigma_x}{E}\,. \tag{B.58}$$

In the case of contraction, we have in addition the following two equations:

$$\varepsilon_y = \frac{1}{E}(-\nu\sigma_x) = -\nu\varepsilon_x\,, \tag{B.59}$$

$$\varepsilon_z = \frac{1}{E}(-\nu\sigma_x) = -\nu\varepsilon_x\,. \tag{B.60}$$

4.4 Uniaxial tensile test and volume change

$$V_e = 3.07328a^3 > V_i = 3a^3 \,. \tag{B.61}$$

4.5 Example: Spherical RVE and pore—definition of damage

$$D = \left(\frac{r_D^3}{R_i^3 + r_D^3}\right)^{\frac{2}{3}} = \left(1 - \frac{\varrho}{\varrho_i}\right)^{\frac{2}{3}} , f = 1 - \frac{\varrho}{\varrho_i} \,. \tag{B.62}$$

4.6 Example: Cubic RVE and spherical pore—definition of damage

$$D = \left(\frac{9\pi}{16}\right)^{\frac{1}{3}} \times \left(1 - \frac{\varrho}{\varrho_i}\right)^{\frac{2}{3}} , f = 1 - \frac{\varrho}{\varrho_i} \,. \tag{B.63}$$

4.7 Cubic RVE and spherical pore—comparison between area and volume damage fraction

$$D = 0.38, \; f = 0.18 \,. \tag{B.64}$$

4.8 Damage evaluation from variation of elastic modulus
The graphical representation is shown in Fig. B.14.

4.9 Volumetric damage measurement based on Archimedes' principle
Under consideration of air buoyancy:

$$\varrho = \frac{\varrho_{fl} - \varrho_{air}}{F_G^{air} - F_G^{fl}} \times F_G^{air} + \varrho_{air} \,. \tag{B.65}$$

Neglecting of air buoyancy:

$$\varrho = \frac{\varrho_{fl}}{F_G^{air} - F_G^{fl}} \times F_G^{air} \,. \tag{B.66}$$

For further details see Ref. [55] (Table B.1).

4.10 Gurson's law: Difference between the cylindrical and spherical pore assumption

4.11 Gurson's law: Expression for volumetric plastic strain
Consider

$$d\varepsilon_x^{pl} = d\lambda \left(\frac{3s_x}{k_t^2} + \frac{D}{k_t} \sinh\left(\frac{3\sigma_m}{2k_t}\right)\right) \tag{B.67}$$

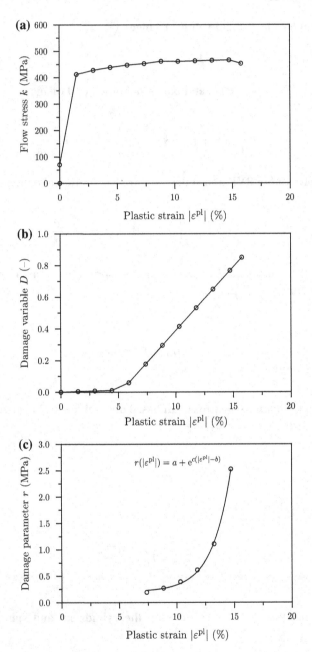

Fig. B.14 **a** Flow curve; **b** damage variable as a function of plastic strain; **c** material parameter r as a function of plastic strain

Table B.1 GURSON law: difference between cylindrical ('c') and spherical ('s') pore assumption

D	Cylindrical pores		Spherical pores		rd_τ	rd_σ
	$\left(\frac{\tau}{k_t}\right)_{x=0}$	$\left(\frac{\sigma}{k_t}\right)_{y=0}$	$\left(\frac{\tau}{k_t}\right)_{x=0}$	$\left(\frac{\sigma}{k_t}\right)_{y=0}$		
0.00	±0.5774	±1.0000	±0.5774	±1.0000	0.0	0.0
0.01	±0.5716	±0.9861	±0.5716	±0.9887	0.0	0.269
0.10	±0.5196	±0.8666	±0.5196	±0.8888	0.0	2.496
0.30	±0.4041	±0.6310	±0.4041	±0.6749	0.0	6.507

The relative difference (rd) between the cylindrical and the spherical approach, i.e. $rd = \left|\frac{f_c - f_s}{f_s}\right| \times 100$, is normalized with the cylindrical value

and

$$s_x + s_y + s_z = 0 \,. \tag{B.68}$$

Problems from Chapter 5

5.10 Stress state on the edge of a circular hole
Fig. B.15 summarizes the stresses which are acting on the edge of a circular hole.

5.11 Critical stress state in an infinite plate with a circular hole
Table B.2 summarizes the stress states.

5.12 Critical stress state in an infinite plate with two circular holes
Fig. B.16 summarizes the stress concentration in the case of two holes.

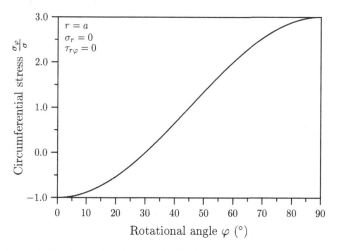

Fig. B.15 Stress distribution on the edge of a circular hole in an infinite plate

Table B.2 Stress states in an infinite plate at different locations

r (mm)	σ_r (MPa)	σ_φ (MPa)	$\tau_{r\varphi}$ (MPa)	σ_{eff} (MPa)
$\varphi = 30°$				
10.0	0.000	0.000	0.000	0.000
12.5	11.664	31.536	−54.622	98.557
15.0	27.778	38.889	−67.358	121.716
17.5	40.820	39.996	−69.275	126.611
20.0	50.625	39.375	−68.120	126.781
22.5	57.956	38.340	−66.407	125.843
$\varphi = 45°$				
10.0	0.000	120.000	0.000	120.000
12.5	21.600	98.400	−63.072	141.272
15.0	33.333	86.667	−77.778	154.536
17.5	40.408	79.592	−79.992	154.750
20.0	45.000	75.000	−78.750	151.260
22.5	48.148	71.852	−76.680	147.177
$\varphi = 90°$				
10.0	0.000	360.000	0.000	360.000
12.5	41.472	232.128	0.000	214.421
15.0	44.444	182.222	0.000	164.565
17.5	39.584	158.784	0.000	143.157
20.0	33.750	146.250	0.000	132.636
22.5	28.532	138.875	0.000	127.035

5.13 Semi-infinite plate with a hole and displacement boundary condition—elasto-plastic material behavior

The results are summarized in Table B.3.

5.14 Semi-infinite plate with a crack and displacement boundary condition

The results are summarized in Table B.4.

5.15 Crack propagating from a circular hole

The results are summarized in Table B.5.

5.16 Crack in a hydraulic cylinder

$$K_{\text{I}} = 13.03 \text{ MPa}\sqrt{\text{m}} < K_{\text{I}}c \quad \text{(overpressurization)}, \tag{B.69}$$

$$K_{\text{I}} = 20.04 \text{ MPa}\sqrt{\text{m}} < K_{\text{I}}c \quad \text{(normal operation)}. \tag{B.70}$$

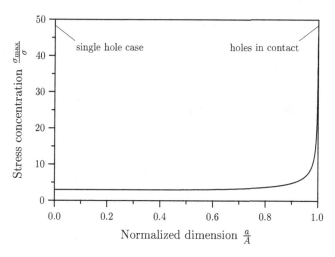

Fig. B.16 Stress concentration in an infinite plate with two holes

Table B.3 Results of plasticity investigation in the thin plate with a circular hole with displacement boundary condition

Material	σ_{nom} (MPa)	σ_{max} (MPa)	Plasticity
7075	197.025	493.745	No
Ti-6Al-4V	305.250	764.957	No
4340	574.425	1439.509	Yes

Table B.4 Results of failure investigation in the thin plate with a crack and displacement boundary condition

Material	σ (MPa)	K_I MPa$\sqrt{m}$	Failure
7075	170.4	30.9	Yes
Ti-6Al-4V	264.0	47.9	No
4340	496.8	90.2	Yes

Table B.5 Results of the failure investigation in the thin plate with a hole and a crack

Case	Quantity	Failure
Only hole	$\sigma_{max} = 450$ MPa	No
Hole + crack	$K_I = 20.0$ MPa$\sqrt{m}$	No
Only crack	$K_I = 4.2$ MPa$\sqrt{m}$	No

5.17 Calibration of an extensometer

$$\varepsilon_{\text{specimen}} = \left| \frac{2}{3} \times \frac{E_2}{E_2} \times \frac{I_2}{I_1} \times \frac{L_1^2}{L_{\frac{h}{2}}^2} \times \varepsilon_{\text{strain gauge}} \right| .$$ (B.71)

5.18 Difference between classical plasticity theory and linear-elastic fracture mechanics

Classical plasticity theory considers the *local* stress, e.g. the maximum stress σ_{max} in the case of the plate with a hole. Linear-elastic fracture mechanics considers the global *stress* σ which is used to calculate the stress intensity factor.

5.19 Elastic strain energy of a uniform bar

$$W_{\text{int}} \int_V w_{\text{int}} \, dV = \int_V \frac{\sigma^2}{2E} dV = \int_0^L \frac{\sigma^2}{2E} A dx = \int_0^L \frac{A}{2E} \left(\frac{F}{A} \right)^2 dx .$$ (B.72)

5.20 Elastic strain energy—Castigliano's theorem

$$u_y(L) = \frac{\partial W_{\text{int}}}{\partial F} = \frac{\partial}{\partial F} \left(\frac{F^2 L^3}{6EI} \right) = \frac{FL^3}{3EI} .$$ (B.73)

5.21 Double cracked bar—energy release rate

$$\mathcal{G}_{\text{II}} = \frac{9F^2}{8Ebh} .$$ (B.74)

5.22 Cracked beam—energy release rate

$$\mathcal{G}_{\text{II}} = \frac{9F^2 a^2}{Eb^2 h^3} .$$ (B.75)

5.23 Standard bend specimen—evaluation of J-integral

$$J = J^{\text{el}} + J^{\text{pl}} = 56.5 + 100 = 156.5 \, \text{kN/m} .$$ (B.76)

References

1. Anderson TL (1995) Fracture mechanics: fundamentals and applications. CRC Press, Boca Raton
2. Asaro RJ, Lubarda VA (2006) Mechanics of solids and materials. Cambridge University Press, Cambridge
3. ASTM (1997) Standard test method for plane-strain fracture toughness of metallic materials. ASTM Standard E 399–90, ASTM, Philadelphia
4. ASTM (2001) Standard test method for measurement of fracture toughness. ASTM Standard E 1820–01, ASTM, Philadelphia
5. Backhaus G (1968) Zur Fließgrenze bei allgemeiner Verfestigung (On the yield condition for general workhardening). Z Angew Math Mech 48:99–108
6. Bauschinger J (1886) Über die Veränderung der Elastizitätsgrenze und die Festigkeit des Eisens und Stahls durch Strecken und Quetschen, durch Erwärmen und Abkühlen und durch oftmals wiederholte Beanspruchungen (On the change of the elastic limit and strength of iron and steel by elongation, shortening, heating, cooling and repeated loading). Mitt Mech-Tech Lab Königl Tech Hochsch München 13:1–116
7. Belytschko T, Liu WK, Moran B (2000) Nonlinear finite elements for continua and structures. Wiley, Chichester
8. Betten J (2001) Kontinuumsmechanik. Springer, Berlin
9. Blumenauer H, Pusch G (1993) Technische Bruchmechanik. Deustcher Verlag fr Grundstoffindustrie, Leipzig
10. de Borst R (1986) Non-linear analysis of frictional materials. Dissertation, Delft University of Technology
11. Bridgman PW (1952) Studies in large plastic flow and fracture. McGraw Hill, New York
12. Broberg KB (1971) Crack-growth criteria and non-linear fracture mechanics. J Mech Phys Solids 19:407–418
13. Callister WD (2001) Fundamentals of materials science and engineering. Wiley, New York
14. Chen WF, Saleeb AF (1982) Constitutive equations for engineering materials. Elasticity and modelling, vol 1. Wiley, New York
15. Chen WF, Han DJ (1988) Plasticity for structural engineers. Springer, New York
16. Davis JR (2004) Tensile Testing. ASM International, Ohio
17. DeHoff RT, Rhines FN (1968) Quantitative microscopy. McGraw-Hill, New York
18. Drucker DC (1952) A more fundamental approach to plastic stress-strain relations. In: Sternberg E et al (eds) Proceedings of 1st U.S. national congress applied mechanics. Edward Brothers Inc, Michigan, pp 487–491

© Springer Science+Business Media Singapore 2016
A. Öchsner, *Continuum Damage and Fracture Mechanics*,
DOI 10.1007/978-981-287-865-6

19. Exner HE, Hougardy HP (1988) Quantitative image analysis of microstructures. DGM Informationsgesellschaft mbH, Oberursel
20. Exner HE (2004) Stereology and 3D microscopy: useful alternatives or competitors in the quantitative analysis of microstructures? Image Anal Stereol 23:73–82
21. Gegner J, Öchsner A, Winter W, Kuhn G (2000) Metallographical investigations of ductile damage in aluminium alloys. Prakt Metallogr-Pr M 37:563–579
22. Gegner J, Öchsner A (2000) Digital image analysis in quantitative metallography. Prakt Metallogr-Pr M 38:499–513
23. Gromada M, Mishuris G, Öchsner A (2011) Correction formulae for the stress distribution in round tensile specimens at neck presence. Springer, Heidelberg
24. Gross D, Hauger W, Schröder J, Wall WA, Bonet J (2011) Engineering mechanics 2: mechanics of materials. Springer, Berlin
25. Gross D, Seelig T (2011) Fracture mechanics: with an introduction to micromechanics. Springer, Berlin
26. Gurson AL (1977) Continuum theory of ductile rupture by void nucleation and growth: part I—yield criteria and flow rules. J Eng Mater Technol ASME 99:2–15
27. Hill R (1950) The mathematical theory of plasticity. Oxford University Press, Oxford
28. Hertzberg RW (1996) Deformation and fracture mechanics of engineering materials. Wiley, Hoboken
29. Lehmann Th (1972) Einige Bemerkungen zu einer allgemeinen Klasse von Stoffgesetzen für große elasto-plastische (Some remarks on a general class of yield conditions for large elasto-plastic deformations) Ing Arch 41:297–310
30. Lemaitre J (1985) A continuous damage mechanics model for ductile fracture. J Eng Mater Technol ASME 107:83–89
31. Lemaitre J (1985) Coupled elasto-plasticity and damage constitutive equations. Comput Method Appl 51:31–49
32. Lemaitre J (1996) A course on damage mechanics. Springer, Berlin
33. Lemaitre J, Desmorat R (2005) Engineering damage mechanics: ductile, creep, fatigue and brittle failures. Springer, Berlin
34. Margolin BZ, Kostylev VI (1999) Analysis of the effect of biaxial loading on the fracture toughness of reactor pressure-vessel steels. Strength Mater 31:433–447
35. Melan E (1938) Zur Plastizität des räumlichen Kontinuums (On the plasticity of the spatial continuum). Ing Arch 9:116–126
36. Moore TA (2013) A general relativity workbook. University Science Books, Mill Valley
37. Murakami Y et al (1987) Stress intensity factors handbook. Pergamon Press, New York
38. Nash WA (1998) Schaum's outline of theory and problems of strength of material. McGraw-Hill, New York
39. Nayak GC, Zienkiewicz OC (1972) Convenient form of stress invariants for plasticity. J Struct Div ASCE 98:1949–1954
40. de Souza Neto EA, Perić D, Owen DRJ (2008) Computational methods for plasticity: theory and applications. Wiley, Chichester
41. Newman JC Jr, Raju IS (1981) An empirical stress-intensity factor equation for the surface crack. Eng Fract Mech 15:185–192
42. Öchsner A, Gegner J, Winter W, Kuhn G (2001) Experimental and numerical investigations of ductile damage in aluminium alloys. Mat Sci Eng A-Struct 318:328–333
43. Öchsner A, Merkel M (2013) One-dimensional finite elements: an introduction to the FE method. Springer, Berlin
44. Öchsner A (2014) Elasto-plasticity of frame structure elements: modelling and simulation of rods and beams. Springer, Berlin
45. Pilkey WD (2005) Formulas for stress, strain, and structural matrices. Wiley, Hoboken
46. Pook L (2007) Metal fatigue—what it is, why it matters. Springer, Dordrecht
47. Prager W (1955) The theory of plasticity; a survey of recent achievements (James Clayton Lecture). Proc Inst Mech Eng

48. Prager W (1956) A new method of analyzing stress and strain in work-hardening plastic solids. J Appl Mech 23:493–496
49. Purushothama Raj P, Ramasamy V (2012) Strength of materials. Pearson, Chennai
50. Radon JC, Leevers PS, Culver LE (1978) Fracture toughness of PMMA under biaxial stress. In: Taplin DMR (ed) Advances in research on the strength and fracture of materials. Pergamon Press, New York
51. Rice JR (1968) A path independent integral and the approximate analysis of strain concentration by notches and cracks. J Appl Mech 35:379–386
52. Rice JR, Paris PC, Merkle JG (1973) Some further results of J-integral analysis and estimates. In: Progress in flaw growth and fracture toughness testing. Special Technical Publication 536, American Society for Testing and Materials, Philadelphia
53. Rice JR, Sorensen EP (1978) Continuing crack-tip deformation and fracture for plane-strain crack growth in elastic-plastic solids. J Mech Phys Solids 26:163–186
54. Rooke DP, Cartwright DJ (1976) Compendium of stress intensity factors. Her Majesty's Station Office, London
55. Sartorius YDK 01, YDK 01-0D, YDK 01LP: Density determination kit users manual. http://www.dcu.ie/sites/default/files/mechanical_engineering/pdfs/manuals/Density%20 Determination%20Kit%28a%29.pdf. Accessed 26 May 2015
56. Schijve J (2001) Fatigue of structures and materials. Kluwer, Dordrecht
57. Shield RT, Ziegler H (1958) On Prager's hardening rule. Z Angew Math Phys 9a:260–276
58. Sih GC (1973) Handbook of stress intensity factors. Institute of Fracture and Solid Mechanics, Lehigh University, Bethlehem
59. Simo JC, Hughes TJR (1998) Computational inelasticity. Springer, New York
60. Sun CT, Jin Z-H (2012) Fracture mechanics. Elsevier, Amsterdam
61. Suzuki H, Ninomiya T, Sumino K, Takeuchi S (1985) Dislocation in solids. In: Proceedings of the IX Yamada. VNU Science Press, Utrecht
62. Tada HP, Paris PC, Irwin GR (1973) The stress analysis of cracks handbook. Del Research Corporation, Hellertown
63. Thomason PF (1990) Ductile fracture of metals. Pergamon Press, Oxford
64. Timoshenko S (1940) Strength of materials—part I elementary theory and problems. D. Van Nostrand Company, New York
65. Timoshenko S, Gere JM (1961) Theory of elastic stability. McGraw-Hill, Aukland
66. Timoshenko SP, Goodier JN (1970) Theory of elasticity. McGraw-Hill, New York
67. Trattnig G, Antretter T, Pippana R (2008) Fracture of austenitic steel subject to a wide range of stress triaxiality ratios and crack deformation modes. Eng Fract Mech 75:223–235
68. Tvergaard V (1991) Mechanical modelling of ductile fracture. Meccanica 26:11–16
69. Underwood EE (1970) Quantitative stereology. Addison-Wesley, Massachusetts
70. Zhu X-K, Joyce JA (2012) Review of fracture toughness (G, K, J, CTOD, CTOA) testing and standardization. Eng Fract Mech 85:1–46
71. Ziegler H (1959) A modification of Prager's hardening rule. Q Appl Math 17:55–65

Index

© Springer Science+Business Media Singapore 2016
A. Öchsner, *Continuum Damage and Fracture Mechanics*,
DOI 10.1007/978-981-287-865-6

Printed in the United States
By Bookmasters